INDUSTRIAL

Discipline-Specific Review for the FE/EIT Exam

University of Missouri–Columbia
Department of Industrial Engineering
James S. Noble, PhD, PE
Thomas J. Crowe, PhD, PE
Elin M. Wicks, PhD
Larry G. David, PhD, PE
Cerry M. Klein, PhD
Luis G. Occena, PhD
Owen M. Miller, DSc, PE
C. Alec Chang, PhD
with Michael R. Lindeburg, PE

Professional Publications, Inc.
Belmont, California

How to Locate Errata and Other Updates for This Book

At Professional Publications, we do our best to bring you error-free books. But when errors do occur, we want to make sure that you know about them so they cause as little confusion as possible.

A current list of known errata and other updates for this book is available on the PPI website at **www.ppi2pass.com**. From the website home page, click on "Errata." We update the errata page as often as necessary, so check in regularly. You will also find instructions for submitting suspected errata. We are grateful to every reader who takes the time to help us improve the quality of our books by pointing out an error.

INDUSTRIAL DISCIPLINE-SPECIFIC REVIEW FOR THE FE/EIT EXAM

Current printing of this edition: 3

Printing History

edition number	printing number	update
1	1	New book.
1	2	Minor corrections.
1	3	Updated front matter.

Printed in the United States of America

Professional Publications, Inc.
1250 Fifth Avenue, Belmont, CA 94002
(650) 593-9119
www.ppi2pass.com

Library of Congress Cataloging-in-Publication Data

Industrial discipline-specific review for the FE/EIT exam / James S. Noble... [et al.].
p. cm.
ISBN 1-888577-22-3
1. Engineering--United States--Examinations--Study guides.
2. Electric engineering--United States--Examinations--Study guides.
3. Engineering--Problems, exercises, etc. 4. Engineers--Certification--United States. I. Noble, James S.
TA159.I48 1997
670'.76--dc21 97-31945
CIP

Table of Contents

Preface and Acknowledgments

This book is one in a series of five that is intended for engineers and students who are taking the engineering discipline-specific (DS) afternoon portion of the Fundamentals of Engineering (FE) exam.

The topics covered in the DS afternoon FE exams are completely different from the topics covered in the morning portion of the FE exam. Since this book only covers one discipline-specific exam, it really addresses only half of the FE exam, and even then, only one specific discipline.

This book consolidates 120 practical review problems, covering all of the discipline-specific exam topics. The practice problems include full solutions. The problems in this book were developed by Robert B. Angus, PE, John E. Hajjar, and Abdulrahman Yassine, following the format, style, subject breakdown, and guidelines that I provided.

In designing this book, I used the NCEES Handbook and the breakdown of problem types published by NCEES. However, as with most standardized tests, there is no guarantee that any specific problem type will be encountered. It is expected that minor variations in problem content will occur from exam to exam.

As with all of Professional Publications' books, the problems in this book are original and have been ethically derived. Although examinee feedback was used to determine its content, this book contains problems that are only *like* those that are on the exam. There are no actual exam problems in this book.

This book was designed to complement my *FE Review Manual*, which you will also need to prepare for the FE exam. The *Review Manual* is Professional Publications' most popular study guide for both the morning and afternoon general exams. It and the *Engineer-In-Training Reference Manual* have been the most popular review books for this exam for more than 20 years.

You cannot prepare adequately without your own copy of the NCEES Handbook. This document contains the data and formulas that you will need to solve both the general and the discipline-specific problems. A good way to become familiar with it is to look up the information, formulas, and data that you need while trying to work practice problems.

No exam-prep book is ever complete. By necessity, it will change as the exam changes. Even when the exam format doesn't change for a while, new problems and improved explanations can always be added. A new edition of this book was already on the design-board as this book was going to the printer.

Michael Lindeburg
mlindeburg@ppi2pass.com

Preface and Acknowledgments

Engineering Registration in the United States

ENGINEERING REGISTRATION

Engineering registration (also known as *engineering licensing*) in the United States is an examination process by which a state's board of engineering licensing (i.e., registration board) determines and certifies that you have achieved a minimum level of competence. This process protects the public by preventing unqualified individuals from offering engineering services.

Most engineers do not need to be registered. In particular, most engineers who work for companies that design and manufacture products are exempt from the licensing requirement. This is known as the *industrial exemption.* Nevertheless, there are many good reasons for registering. For example, you cannot offer consulting engineering design services in any state unless you are registered in that state. Even within a product-oriented corporation, however, you may find that employment, advancement, or managerial positions are limited to registered engineers.

Once you have met the registration requirements, you will be allowed to use the titles Professional Engineer (PE), Registered Engineer (RE), and Consulting Engineer (CE).

Although the registration process is similar in all 50 states, each state has its own registration law. Unless you offer consulting engineering services in more than one state, however, you will not need to register in other states.

The U.S. Registration Procedure

The registration procedure is similar in most states. You will take two eight-hour written examinations. The first is the *Fundamentals of Engineering Examination*, also known as the *Engineer-In-Training Examination* and the *Intern Engineer Exam.* The initials FE, EIT, and IE are also used. This examination covers basic subjects from all of the mathematics, physics, chemistry, and engineering classes you took during your first four university years.

In rare cases, you may be allowed to skip this first examination. However, the actual details of registration qualifications, experience requirements, minimum education levels, fees, oral interviews, and examination schedules vary from state to state. Contact your state's registration board for more information.

The second eight-hour examination is the *Principles and Practice of Engineering Exam.* The initials PE are also used. This examination covers subjects only from your areas of specialty.

National Council of Examiners for Engineering and Surveying

The National Council of Examiners for Engineering and Surveying (NCEES) in Clemson, South Carolina, produces, distributes, and scores the national FE and PE examinations. The individual states purchase the examinations from NCEES and administer them themselves. NCEES does not distribute applications to take the examinations, administer the examinations or appeals, or notify you of the results. These tasks are all performed by the states.

Reciprocity Among States

With minor exceptions, having a license from one state will not permit you to practice engineering in another state. You must have a professional engineering license from each state in which you work. For most engineers, this is not a problem, but for some, it is. Luckily, it is not too difficult to get a license from every state you work in once you have a license from one state.

All states use the NCEES examinations. If you take and pass the FE or PE examination in one state, your certificate will be honored by all of the other states. Although there may be other special requirements imposed by a state, it will not be necessary to retake the FE and PE examinations. The issuance of an engineering license based on another state's license is known as *reciprocity* or *comity.*

The simultaneous administration of identical examinations in all states has led to the term *uniform examination.* However, each state is still free to choose its own minimum passing score and to add special questions and requirements to the examination process. Therefore, the use of a uniform examination has not, by itself, ensured reciprocity among states.

THE FE EXAMINATION

Applying for the Examination

Each state charges different fees, specifies different requirements, and uses different forms to apply for the exam. Therefore, it will be necessary to request an application from the state in which you want to become registered. Generally, it is sufficient for you to phone for this application. You'll find contact information (websites, telephone numbers, email addresses, etc.) for all U.S. state and territorial boards of registration at www.ppi2pass.com. Click on the State Boards link.

Keep a copy of your examination application and send the original application by certified mail, requesting a receipt of delivery. Keep your proof of mailing and delivery with your copy of the application.

Examination Dates

The national FE and PE examinations are administered twice a year (usually in mid-April and late October), on the same weekends in all states. Check www.ppi2pass.com for a current exam schedule. Click on the Exam FAQs link.

FE Examination Format

The NCEES Fundamentals of Engineering examination has the following format and characteristics.

- There are two four-hour sessions separated by a one-hour lunch.
- Examination questions are distributed in a bound examination booklet. A different examination booklet is used for each of these two sessions.
- The morning session (also known as the *A.M. session*) has 120 multiple-choice questions, each with four possible answers lettered (A) through (D). Responses must be recorded with a pencil provided by NCEES on special answer sheets. No credit is given for answers recorded in ink.
- Each problem in the morning session is worth one point. The total score possible in the morning is 120 points. Guessing is valid; no points are subtracted for incorrect answers.
- There are questions on the examination from most of the undergraduate engineering degree program subjects. Questions from the same subject are all grouped together, and the subjects are labeled. The percentages of questions for each subject in the morning session are given in the following table.

Morning FE Exam Subjects

subject	percentage of questions (%)
chemistry	9
computers	6
dynamics	7
electrical circuits	10
engineering economics	4
ethics	4
fluid mechanics	7
materials science and structure of matter	7
mathematics	20
mechanics of materials	7
statics	10
thermodynamics	9

- There are seven different versions of the afternoon session (also known as the *P.M. session*), six of which correspond to a specific engineering discipline: chemical, civil, electrical, environmental, industrial, and mechanical engineering.

Each version of the afternoon session consists of 60 questions. All questions are mandatory. Questions in each subject may be grouped into related problem sets containing between two and ten questions each.

The seventh version of the afternoon examination is a general examination suitable for anyone, but in particular, for engineers whose specialties are not one of the other six disciplines. Though the subjects in the general afternoon examination correspond to the morning subjects, the questions are more complex—hence their double weighting. Questions on the afternoon examination are intended to cover concepts learned in the last two years of a four-year degree program. Unlike morning questions, these questions may deal with more than one basic concept per question.

The percentages of questions for each subject in the general afternoon session examination are given in the following table.

Afternoon FE Exam Subjects (General Exam)

subject	percentage of questions (%)
chemistry	7.5
computers	5
dynamics	7.5
electrical circuits	10
engineering economics	5
ethics	5
fluid mechanics	7.5
materials science and structure of matter	5
mathematics	20
mechanics of materials	7.5
statics	10
thermodynamics	10

The percentages of questions for each subject in the industrial discipline-specific afternoon session examination are as follows. The discipline-specific afternoon examinations cover substantially different bodies of knowledge than the morning examination. Formulas and tables of data needed to solve questions in these examinations will be included in either the NCEES Handbook or in the body of the question statement itself.

Afternoon FE Exam Subjects (DS Exam)

subject	percentage of questions (%)
chemistry	11
computers	7
dynamics	9
electrical circuits	12
engineering economics	5
ethics	5
fluid mechanics	8
materials science and structure of matter	8
mathematics	24
mechanics of materials	8
statics	12
thermodynamics	11

Each afternoon question consists of a problem statement followed by multiple-choice questions. Four answer choices lettered (A) through (D) are given, from which you must choose the best answer.

- Each question in the afternoon is worth two points, making the total possible score 120 points.
- The scores from the morning and afternoon sessions are added together to determine your total score. No points are subtracted for guessing or incorrect answers. Both sessions are given equal weight. It is not necessary to achieve any minimum score on either the morning or afternoon sessions.
- All grading is done by computer optical sensing.

Use of SI Units on the FE Exam

Metric questions are used in all subjects, except some civil engineering and surveying subjects that typically use only customary U.S. (i.e., English) units. SI units are consistent with ANSI/IEEE standard 268 (the American Standard for Metric Practice). Non-SI metric units might still be used when common or where needed for consistency with tabulated data (e.g., use of bars in pressure measurement).

Grading and Scoring the FE Exam

The FE exam is not graded on the curve, and there is no guarantee that a certain percent of examinees will pass. Rather, NCEES uses a modification of the Angoff procedure to determine the suggested passing score (the cutoff point or cut score).

With this method, a group of engineering professors and other experts estimate the fraction of minimally qualified engineers that will be able to answer each question correctly. The summation of the estimated fractions for all test questions becomes the passing score. The passing score in recent years has been somewhat less than 50 percent (i.e., a raw score of approximately 110 points out of 240). Because the law in most states requires engineers to achieve a score of 70 percent to become licensed, you may be reported as having achieved a score of 70 percent if your raw score is greater than the passing score established by NCEES, regardless of the raw percentage. The actual score may be slightly more or slightly less than 110 as determined from the performance of all examinees on the equating subtest.

Approximately 20 percent of each FE exam consists of questions repeated from previous examinations—this is the *equating subtest.* Since the performance of previous examinees on the equating subtest is known, comparisons can be made between the two examinations and examinee populations. These comparisons are used to adjust the passing score.

The individual states are free to adopt their own passing score, but all adopt NCEES's suggested passing score because the states believe this cutoff score can be defended if challenged.

You will receive the results approximately 12 to 14 weeks after the examination. If you pass, your score may or may not be revealed to you, depending on your state's policy, but if you fail, you will receive your score.

The following table lists the approximate fractions of examinees passing the FE exam.

Approximate FE Exam Passing Rates

category	percent passing
total, all U.S. states	60%–70%
ABET accredited, four-year engineering degrees[a]	70%–80%
nonaccredited, four-year engineering degrees	50%–65%
ABET accredited, four-year technology degrees[a]	35%–45%
nonaccredited, four-year technology degrees	25%–35%
nongraduates	35%–40%

[a] The Accreditation Board for Engineering and Technology (ABET) reviews and approves engineering degree programs in the United States. No engineering degree programs offered by universities outside of the United States and its territories or the Commonwealth of Puerto Rico are accredited by ABET.

Permitted Reference Material

Since October 1993, the FE examination has been what NCEES calls a "limited-reference" exam. This means that no books or references other than those supplied by NCEES may be used. Therefore, the FE examination is really an "NCEES-publication only" exam. NCEES provides its own FE Reference Handbook for use during the examination. No books from other publishers may be used.

CALCULATORS

In most states, battery- or solar-powered, silent calculators can be used, although printers cannot be used. (The solar-powered calculators are preferred because they do not have batteries that run down.) In most states, programmable, preprogrammed, or business/finance calculators are allowed. Similarly, nomographs and specialty slide rules are permitted. To prevent unauthorized transcription and redistribution of the examination questions, calculators with communication or text editing capabilities are banned from all NCEES exam sites. You cannot share calculators with other examinees.

It is essential that a calculator used for engineering examinations have the following functions.

- trigonometric functions
- inverse trigonometric functions
- hyperbolic functions
- pi
- square root and x^2
- common and natural logarithms
- y^x and e^x

For maximum speed, your calculator should also have or be programmed for the following functions.

- extracting roots of quadratic and higher-order equations
- converting between polar (phasor) and rectangular vectors
- finding standard deviations and variances
- calculating determinants of 3×3 matrices
- linear regression
- economic analysis and other financial functions

STRATEGIES FOR PASSING THE FE EXAM

The most successful strategy to pass the FE exam is to prepare in all of the examination subjects. Do not limit the number of subjects you study in hopes of finding enough questions in your particular areas of knowledge to pass.

Fast recall and stamina are essential to doing well. You must be able to quickly recall solution procedures, formulas, and important data. You will not have time during the exam to derive solutions methods—you must know them instinctively. This ability must be maintained for eight hours. Be sure to gain familiarity with the NCEES Handbook by using it as your only reference for some of the problems you work when you study.

In order to get exposure to all examination subjects, it is imperative that you develop and adhere to a review schedule. If you are not taking a classroom review course (where the order of your preparation is determined by the lectures), prepare your own review schedule.

There are also physical demands on your body during the examination. It is very difficult to remain alert and attentive for eight hours or more. Unfortunately, the more time you study, the less time you have to maintain your physical condition. Thus, most examinees arrive at the examination site in peak mental condition but in deteriorated physical condition. While preparing for the FE exam is not the only good reason for embarking on a physical conditioning program, it can serve as a good incentive to get in shape.

It will be helpful to make a few simple decisions prior to starting your review. You should be aware of the different options available to you. For example, you should decide early on to

- use SI units in your preparation
- perform electrical calculations with effective (rms) or maximum values
- take calculations out to a maximum of four significant digits
- prepare in all examination subjects, not just your specialty areas

At the beginning of your review program, you should locate a spare calculator. It is not necessary to buy a spare if you can arrange to borrow one from a friend or the office. However, if possible, your primary and spare calculators should be identical. If your spare calculator is not identical to the primary calculator, spend some time familiarizing yourself with its functions.

A Few Days Before the Exam

There are a few things you should do a week or so before the examination date. For example, visit the exam site in order to find the building, parking areas, examination room, and rest rooms. You should also make arrangements for child care and transportation. Since the examination does not always start or end at the designated times, make sure that your child care and transportation arrangements can tolerate a later-than-scheduled completion.

Next in importance to your scholastic preparation is the preparation of your two examination kits. The first kit consists of a bag or box containing items to bring with you into the examination room.

- [ ] letter admitting you to the examination
- [ ] photographic identification
- [ ] main calculator
- [ ] spare calculator
- [ ] extra calculator batteries
- [ ] a large eraser
- [ ] unobtrusive snacks
- [ ] travel pack of tissues
- [ ] headache remedy
- [ ] $2.00 in change
- [ ] light, comfortable sweater
- [ ] loose shoes or slippers
- [ ] handkerchief
- [ ] cushion for your chair
- [ ] small hand towel
- [ ] earplugs
- [ ] wristwatch with alarm
- [ ] wire coat hanger
- [ ] extra set of car keys

The second kit consists of the following items and should be left in a separate bag or box in your car in case they are needed.

- [ ] copy of your application
- [ ] proof of delivery
- [ ] this book
- [ ] other references
- [ ] regular dictionary
- [ ] scientific dictionary
- [ ] course notes in three-ring binders
- [ ] cardboard box (use as a bookcase)
- [ ] instruction booklets for all your calculators
- [ ] light lunch
- [ ] beverages in thermos and cans
- [ ] sunglasses
- [ ] extra pair of prescription glasses
- [ ] raincoat, boots, gloves, hat, and umbrella
- [ ] street map of the examination site
- [ ] note to the parking patrol for your windshield explaining where you are, what you are doing, and why your time may have expired
- [ ] battery-powered desk lamp

The Day Before the Exam

Take the day before the examination off from work to relax. Do not cram the last night. A good prior night's sleep is the best way to start the examination. If you live far from the examination site, consider getting a hotel room in which to spend the night.

Make sure your exam kits are packed and ready to go.

The Day of the Exam

You should arrive at least 30 minutes before the examination starts. This will allow time for finding a convenient parking place, bringing your materials to the examination room, and making room and seating changes. Be prepared, though, to find that the examination room is not open or ready at the designated time.

Once the examination has started, consider the following suggestions.

- Set your wristwatch alarm for five minutes before the end of each four-hour session and use that remaining time to guess at all of the remaining unsolved problems. Do not work up until the very end. You will be successful with about 25 percent of your guesses, and these points will more than make up for the few points you might earn by working during the last five minutes.
- Do not spend more than two minutes per morning question. (The average time available per problem is two minutes.) If you have not finished a question in that time, make a note of it and continue on.
- Do not ask your proctors technical questions. Even if they are knowledgeable in engineering, they will not be permitted to answer your questions.
- Make a quick mental note about any problems for which you cannot find a correct response or for which you believe there are two correct answers. Errors in the exam are rare, but they do occur. Being able to point out an error later might give you the margin you need to pass. Since such problems are almost always discovered during the scoring process and discounted from the examination, it is not necessary to tell your proctor, but be sure to mark the one best answer before moving on.
- Make sure all of your responses on the answer sheet are dark and completely fill the bubbles.

Common Questions About the DS Exam

Q: Do I have to take the DS exam?

A: Most people do not have to take the DS exam and may elect the general exam option. The state boards do not care which afternoon option you choose; nor do employers. In some cases, examinees who are still in their undergraduate degree program may be required by their university to take a specific DS exam.

Q: Do all mechanical, civil, electrical, chemical, industrial, and environmental engineers take the DS exam?

A: Originally, the concept was that examinees from the "big five" disciplines would take the DS exam, and the general exam would be for everyone else. This remains just a concept, however. A majority of engineers in all of the disciplines apparently take the general exam.

Q: When do I elect to take the DS exam?

A: You will make your decision on the afternoon of the FE exam, when the exam booklet (containing all of the DS exams) is distributed to you.

Q: Where on the application for the FE exam do I choose which DS exam I want to take?

A: You don't specify the DS option at the time of your application.

Q: After starting to work on either the DS or general exam, can I change my mind and switch options?

A: Yes. Theoretically, if you haven't spent too much time on one exam, you can change your mind and start a different one. (You might need to obtain a new answer sheet from the proctor.)

Q: After I take the DS exam, does anyone know that I took it?

A: After you take the FE exam, only NCEES and your state board will know whether you took the DS or general exam. Such information may or may not be retained by your state board.

Q: Will my DS FE certificate be recognized by other states?

A: Yes. All states recognize passing the FE exam and do not distinguish between the DS and general afternoon portions of the FE exam.

Q: Is the DS FE certificate "better" than the general FE certificate?

A: There is no difference. No one will know which option you chose. It's not stated on the certificate you receive from your state.

Q: What is the format of the DS exam?

A: The DS exam is 4 hours long. There are 60 problems, each worth 2 points. The average time per problem is 4 minutes. Each problem is multiple choice with 4 answer choices. Most problems require the application of more than one concept (i.e., formula).

Q: Is there anything special about the way the DS exam is administered?

A: In all ways, the DS and general afternoon exam are equivalent. There is no penalty for guessing. No credit is given for scratch pad work, methods, etc.

Q: Are the answer choices close or tricky?

A: Answer choices are not particularly close together in value, so the number of significant digits is not going to be an issue. Wrong answers, referred to as "distractors" by NCEES, are credible. However, the exam is not "tricky"; it does not try to mislead you.

Q: Are any problems in the afternoon related to each other?

A: Several questions may refer to the same situation or figure. However, NCEES has tried to make all of the questions independent. If you make a mistake on one question, it shouldn't carry over to another.

Q: Is there any minimum passing score for the DS exam?

A: No. It is the total score from your morning and afternoon sessions that determines your passing, not the individual session scores. You do not have to "pass" each session individually.

Q: Is the general portion easier, harder, or the same as the DS exams?

A: Theoretically, all of the afternoon options are the same. At least, that is the intent of offering the specific options: to reduce the variability. Individual passing rates, however, may still vary 5 to 10 percent from exam to exam. (Professional Publications lists the most recent passing statistics for the various DS options on its website at www.ppi2pass.com.)

Q: Do the DS exams cover material at the undergraduate or graduate level?

A: Like the general exam, test topics come entirely from the typical undergraduate degree program. However, the emphasis is primarily on material from the third and fourth year of your program. This may put examinees who take the exam in their junior year at a disadvantage.

Q: Do you need practical work experience to take the DS exam?

A: No.

Q: Does the DS exam also draw on subjects that are in the general exam?

A: Yes. The dividing line between general and DS topics is often indistinct.

Q: Is the DS exam in customary U.S. or SI units?

A: The DS exam is essentially entirely in SI units. A few exceptions exist for some civil engineering subjects (surveying, hydrology, code-based design, etc.) where current common practice is limited to customary U.S. units.

Q: Does the NCEES Handbook cover everything that is on the DS exam?

A: No. You may be tested on subjects that are not present in the NCEES Handbook. However, NCEES has apparently adopted an unofficial policy of providing any necessary information, data, and formulas in the stem of the question. You will not be required to memorize any formulas.

Q: How is the DS reference material identified in the NCEES Handbook?

A: In most cases, the DS reference material is consolidated in the back. However, this policy is not consistently followed. In some cases, the DS reference material is mixed in with the general reference material. Only in some cases is this "intermixed" material identified as being DS material.

Q: Is everything in the DS portion of the NCEES Handbook going to be on the exam?

A: Apparently, there is a fair amount of reference material that isn't needed for every exam. There is no way, however, to know what material is needed.

Q: How long does it take to prepare for the DS exam?

A: Preparing for the DS exam is similar to preparing for a mini PE exam. Engineers typically take two to three months to complete a thorough review for the PE exam. However, examinees who are still in their degree program at a university probably aren't going to spend more than two weeks thinking about, worrying about, or preparing for the DS exam. They rely on their recent familiarity with the subject matter.

Q: If I take the DS exam and fail, do I have to take the DS exam the next time?

A: No. The examination process has no memory.

Q: Where can I get even more information about the DS exam?

A: If you have internet access, visit the Exam FAQs and the Engineering Exam Forum at Professional Publications' website. Our address is www.ppi2pass.com.

How to Use this Book

HOW EXAMINEES CAN USE THIS BOOK

This book is divided into two parts: The first part consists of 60 representative practice problems covering all of the topics in the afternoon DS exam. 60 problems happen to correspond to the number of problems in the afternoon DS exam. You may time yourself by allowing approximately 4 minutes per problem when attempting to solve these problems, but that was not my intent when designing this book. Since the solution follows directly after each problem in this section, I intended for you to read through the problems, attempt to solve them on your own, become familiar with the support material in the official NCEES Handbook, and accumulate the reference materials you think you will need for additional study.

The second part of this book consists of a complete sample examination that you can use as a source of additional practice problems or as a timed diagnostic tool. It also contains 60 problems, and the number of problems in each subject corresponds to the breakdown of subjects published by NCEES. Since the solutions to this part of the book are consolidated at the end, it was my intent that you would solve these problems in a realistic mock-exam mode.

You should use the NCEES Handbook as your only reference during this mock exam.

The morning general exam and the afternoon DS exam essentially cover two different bodies of knowledge. It takes a lot of discipline to prepare for two standardized exams simultaneously. Because of that (and because of my good understanding of human nature), I suspect that you will be tempted to start preparing for your chosen DS exam only after you have become comfortable with the general subjects. That's actually quite logical, because if you run out of time, you will still have the general afternoon exam as a viable option.

If, however, you are limited in time to only two or three months of study, it will be quite difficult to do a thorough DS review if you wait until after you have finished your general review. With a limited amount of time, you really need to prepare for both exams in parallel.

HOW INSTRUCTORS CAN USE THIS BOOK

The availability of the discipline-specific FE exam has greatly complicated the lives of review course instructors and coordinators. The general consensus is that it is essentially impossible to do justice to all of the general FE exam topics and then present a credible review for each of the DS topics. Increases in course cost, expenses, course length, and instructor pools (among many other issues) all conspire to create quite a difficult situation.

One-day reviews for each DS subject are subject-overload from a reviewing examinee's standpoint. Efforts to shuffle FE students over the parallel PE review courses meet with scheduling conflicts. Another idea, that of lengthening lectures and providing more in-depth coverage of existing topics (e.g., covering transistors during the electricity lecture), is perceived as a misuse of time by a majority of the review course attendees. Is it any wonder that virtually every FE review course in the country has elected to only present reviews for the general afternoon exam?

But, while more than half of the examinees elect to take the general afternoon exam, some may actually be required to take a DS exam. This is particularly the case in some university environments where the FE exam has become useful as an "outcome assessment tool." Thus, some method of review is still needed.

Since most examinees begin reviewing approximately two to three months before the exam (which corresponds to when most review courses begin), it is impractical to wait until the end of the general review to start the DS review. The DS review must proceed in parallel with the general review.

In the absence of parallel DS lectures (something that isn't yet occurring in too many review courses), you may want to structure your review course to provide lectures only on the general subjects. Your DS review could be assigned as "independent study," using chapters and problems from this book. Thus, your DS review would consist of distributing this book with a schedule of assignments. Your instructional staff could still provide assistance on specific DS problems, and completed DS assignments could still be recorded.

The final chapter on incorporating DS subjects into review courses has yet to be written. Like the landscape architect who waits until a well-worn path appears through the plants before placing stepping stones, we need to see how review courses do it before we can give any advice.

Practice Problems

COMPUTER COMPUTATIONS AND MODELING

1. The hexadecimal number 7E16 is equivalent to which binary number?

(A) 0111 1011 0001 0101
(B) 1000 0100 1110 1010
(C) 0111 1110 0001 0110
(D) 1000 1110 0101 0101

Solution:

The conversion is made by realizing that each set of 4 binary digits (bits) represents one hexadecimal digit.

$$\begin{aligned} 7_{16} &= 0111_2 \\ \mathrm{E}_{16} &= 1110_2 \\ 1_{16} &= 0001_2 \\ 6_{16} &= 0110_2 \\ 7\mathrm{E}16_{16} &= 0111\ 1110\ 0001\ 0110 \end{aligned}$$

Answer is C.

2. An industrial engineer has been asked to write a computer code that will determine the mean and standard deviation for the process time of a given part. The data input file to be read contains information on all parts produced in the plant. Data that is corrupted is not retained for calculation. Which of the following flow charts represents the correct logic for the computer code?

(A)

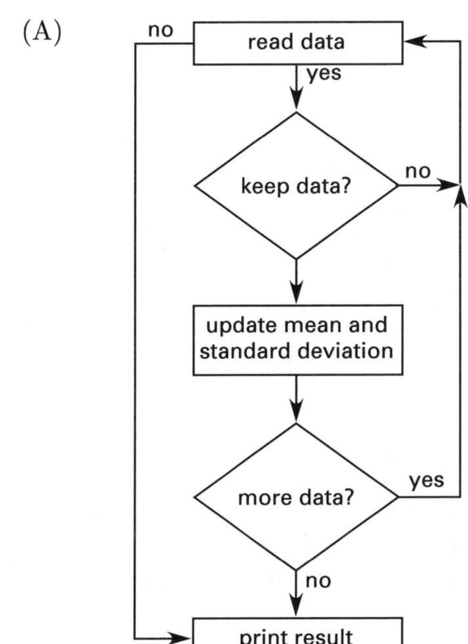

(B)

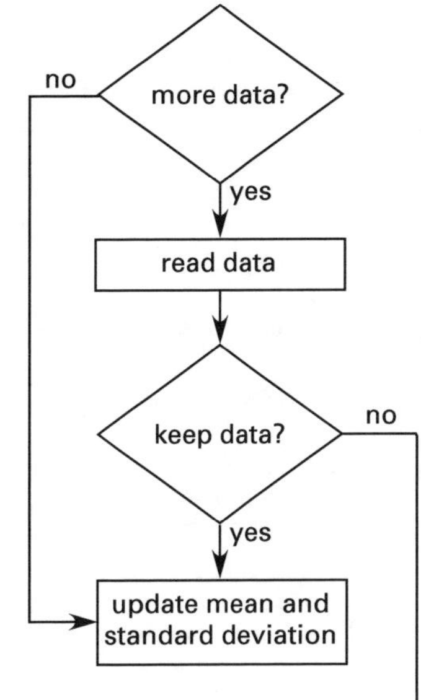

(C)

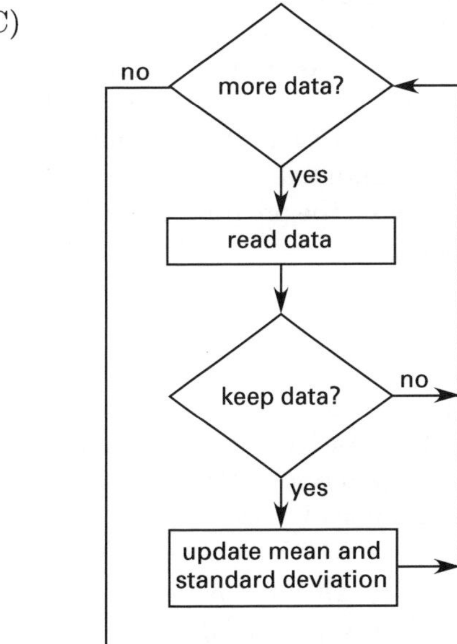

(D)

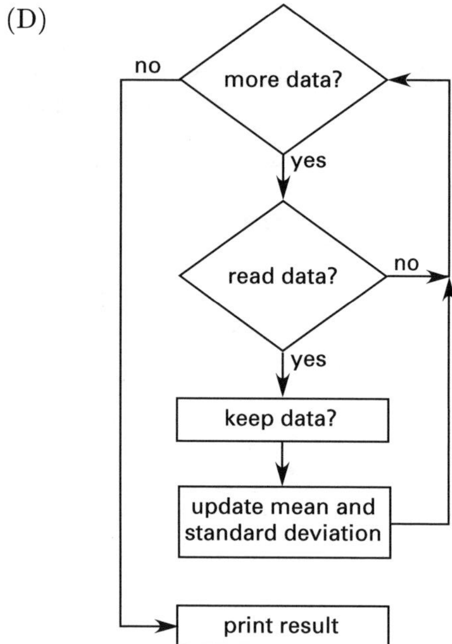

Solution:

First, the program must check to see if there is more data to be read. If there is, then the data is read, otherwise the current result is printed. If data is read, then the program must determine if the data is the correct type. If it is the correct type, then the mean and standard deviation are updated and the program looks for more data. Otherwise, the program looks for more data and skips updating.

Answer is C.

3. Computer programming languages have commands that can be used to test whether given conditions have been satisfied. Let p and q be two conditions. Some of the commands in particular are

$$-p = \text{not } p$$
$$p + q = p \text{ or } q$$
$$p \cdot q = p \text{ and } q$$

The truth table for these commands is

p	q	$-p$	$p+q$	$p \cdot q$
T	T	F	T	T
T	F	F	T	F
F	T	T	T	F
F	F	T	F	F

Give the truth table for the following command.

$$-(p + (p \cdot q)) + (p \cdot (p \cdot -q))$$

(A)

p	q	$-(p + (p \cdot q)) + (p \cdot (p \cdot -q))$
T	T	F
T	F	F
F	T	F
F	F	T

(B)

p	q	$-(p + (p \cdot q)) + (p \cdot (p \cdot -q))$
T	T	T
T	F	T
F	T	F
F	F	F

(C)

p	q	$-(p + (p \cdot q)) + (p \cdot (p \cdot -q))$
T	T	F
T	F	T
F	T	T
F	F	T

(D)

p	q	$-(p + (p \cdot q)) + (p \cdot (p \cdot -q))$
T	T	T
T	F	F
F	T	T
F	F	F

Solution:

To find the correct truth table, start inside the parentheses and work out. Starting on the left side of the command, obtain the following intial truth table.

p	q	$p \cdot q$	$(p + (p \cdot q))$	$-(p + (p \cdot q))$
T	T	T	T	F
T	F	F	T	F
F	T	F	F	T
F	F	F	F	T

The right side of the command yields the following.

p	q	$-q$	$p \cdot -q$	$(p \cdot (p \cdot -q))$
T	T	F	F	F
T	F	T	T	T
F	T	F	F	F
F	F	T	F	F

p	q	$-(p + (p \cdot q))$	$(p \cdot (p \cdot -q))$	$-(p + (p \cdot q)) + (p \cdot (p \cdot -q))$
T	T	F	F	F
T	F	F	T	T
F	T	T	F	T
F	F	T	F	T

Answer is C.

DESIGN OF INDUSTRIAL EXPERIMENTS

4. A comparative study is made for two heat treatment processes. The results of hardness testing (in Rockwell B) of ten sample pairs are shown as follows.

specimen	1	2	3	4	5	6	7	8	9	10
process 1	63	62	69	65	64	67	63	63	64	68
process 2	62	64	69	64	65	66	63	65	63	68

The differences in the individual pairs, d_i, are normally distributed. Calculate the test statistics and conclude whether these two processes produce different results based on a 0.05 significance level.

(A) $t_0 = -0.26$, so there is no evidence to indicate that these two processes produce different hardness readings.

(B) $t_0 = -0.10$, so there is no evidence to indicate that these two processes produce different hardness readings.

(C) $t_0 = 1.62$, so there is significant evidence to indicate that these two processes produce different hardness readings.

(D) $t_0 = 2.26$, so there is significant evidence to indicate that these two processes produce different hardness readings.

Solution:

The pair differences ($d_i = x_{1,i} - x_{2,i}$) are

$$d_i\text{: } 1, -2, 0, 1, -1, 1, 0, -2, 1, 0$$

$$\sum d_i = -1$$

$$\sum d_i^2 = 13$$

$$n = 10$$

The mean difference is

$$\bar{d} = \frac{1}{n}\sum d_i = -\frac{1}{10} = -0.10$$

The estimate of standard deviation is

$$s_d = \sqrt{\frac{\sum d_i^2 - \frac{1}{n}\left(\sum d_i\right)^2}{n-1}} = \sqrt{\frac{13 - \left(\frac{1}{10}\right)(-1)^2}{10-1}}$$

$$= 1.20$$

$$t_0 = \frac{\bar{d}}{\frac{s_d}{\sqrt{n}}} = \frac{-0.10}{\frac{1.20}{\sqrt{10}}} = -0.26$$

From the t-distribution table with $(n-1)$ degrees of freedom, $t^*_{9,1-0.025} = -2.262$. There is no evidence to indicate that these two processes produce different hardness readings.

Answer is A.

5. A one-way factorial design is used to determine whether the sample effect is significant at a 0.05 level. The analysis of the variance table from experimental results is shown as follows.

source of variation	degrees of freedom	sum of squares
between samples	1	4623
within samples	20	8120

The conclusions are

(A) $F = 4061/4623 = 0.88$, so the sample effect is not significant at an 0.05 level.

(B) $F = 4623/8120 = 0.57$, so the sample effect is not significant at an 0.05 level.

(C) $F = 8120/4623 = 1.76$, so the sample effect is significant at an 0.05 level.

(D) $F = 4623/406 = 11.39$, so the sample effect is significant at an 0.05 level.

Solution:

From the table, $F^*_{1,20,0.05} = 4.35$.

$$F = \frac{\text{mean of squares between samples}}{\text{mean of squares within samples}} = \frac{\frac{4623}{1}}{\frac{8120}{20}}$$

$$= 11.39$$

The sample effect is significant at an 0.05 level; that is, the differences between the samples are more significant than the fluctuation within the samples.

Answer is D.

6. A two-way factorial design is used to determine whether the treatment is significant at an 0.05 level. The analysis of the variance table from experimental results is shown as follows.

source of variation	degrees of freedom	sum of squares
replications	4	18.56
treatments	4	36.92
residuals	16	116.08

The conclusions are

(A) $F = 116.08/36.92 = 3.14$, so the treatment is significant at an 0.05 level.
(B) $F = 116.08/18.56 = 6.29$, so the treatment is significant at an 0.05 level.
(C) $F = 7.255/9.23 = 0.786$, so the treatment is not significant at an 0.05 level.
(D) $F = 9.23/7.255 = 1.27$, so the treatment is not significant at an 0.05 level.

Solution:

From the table, $F^*_{4,16,0.05} = 3.01$.

$$F = \frac{\text{mean of squares between samples}}{\text{mean of squares within samples}} = \frac{\frac{36.92}{4}}{\frac{116.08}{16}}$$

$$= 1.27$$

The sample effect from treatment is not significant at an 0.05 level; that is, the fluctuation within the samples is more significant than the differences between samples.

Answer is D.

ENGINEERING ECONOMICS

7. Kelly just won the state lottery. The \$2,000,000 jackpot will be paid in 20 annual installments of \$100,000. The first installment is given to Kelly immediately. The interest rate is 6% compounded yearly. The present worth of Kelly's lottery winnings (at the time of receiving the first installment) is most nearly

(A) \$1,115,810
(B) \$1,146,990
(C) \$1,215,810
(D) \$1,900,000

Solution:

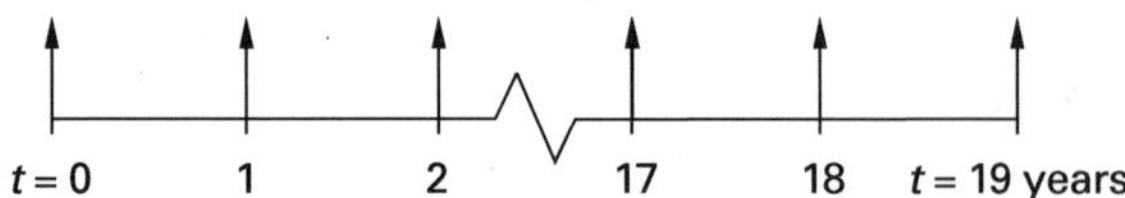

Use the uniform series present worth factor to find the present worth of the last 19 installments at the time the first installment is made. Add this value to the first installment to determine the total present worth of Kelly's lottery winnings.

$$P = \$100{,}000 + A(P/A, i\%, n)$$

A = uniform series of end of compounding period cash flows $= \$100{,}000$

i = interest rate per compounding period = 6%/year

n = number of compounding periods = 19 years

P = present equivalent value of a cash flow or series of cash flows

$$= \$100{,}000 + (\$100{,}000)(P/A, 6\%, 19)$$

$$= \$100{,}000 + (\$100{,}000)\left[\frac{(1+0.06)^{19}-1}{(0.06)(1+0.06)^{19}}\right]$$

$$= \$100{,}000 + (\$100{,}000)(11.1581)$$

$$= \$1{,}215{,}810$$

Answer is C.

8. Joe wants to be a millionaire. To achieve this goal, he invests \$5000 each year into an account that pays 10% interest, compounded yearly. The amount of time it will take Joe to reach his goal of becoming a millionaire is

(A) 20 yr
(B) 32 yr
(C) 43 yr
(D) 181 yr

Solution:

Use the uniform series compound amount factor to relate the future worth of the account to the yearly deposits.

$$F = A(F/A, i\%, n)$$

A = uniform series of end of compounding period cash flows $= \$5000$

i = interest rate per compounding period = 10%/year

n = number of compounding periods = number of years

F = future equivalent value of a cash flow or series of cash flows $= \$1{,}000{,}000$

$$\$1{,}000{,}000 = (\$5000)(F/A, 10\%, n)$$

$$= (\$5000)\left[\frac{(1+0.1)^n - 1}{0.1}\right]$$

$$21 = 1.1^n$$

$$\log(21) = n\log(1.1)$$

$$n = \frac{\log(21)}{\log(1.1)} = 31.94 \text{ yr} \quad (32 \text{ yr})$$

Answer is B.

9. A credit card company offers students a credit line of \$2000 and charges an annual percentage rate of 12%, compounded daily. What is the effective annual interest rate?

(A) 3.28%
(B) 12.00%
(C) 12.75%
(D) 13.19%

Solution:

The following equation relates a nominal interest rate to the effective annual interest rate.

$$i_{\text{eff}} = \left(1 + \frac{r}{M}\right)^M - 1$$

$$\begin{aligned} r &= \text{nominal interest rate} = 12\% \\ M &= \text{number of compounding periods per year} \\ &= 365 \text{ compounding periods per year} \\ i_{\text{eff}} &= \text{effective annual interest rate} \\ &= \left(1 + \frac{0.12}{365}\right)^{365} - 1 = 0.1275 \quad (12.75\%) \end{aligned}$$

Answer is C.

ENGINEERING STATISTICS

10. Defects in the finished surface of furniture are approximately Poisson distributed with a mean of 0.015 defects/m^2. A finished bookshelf board containing 10 m^2 of finished surface is considered defective if it contains more than one defect. If shelves are packaged 50 per box, what is the probability that a box contains fewer than 2 defective items?

(A) 0.50
(B) 0.61
(C) 0.82
(D) 0.91

Solution:

Let x = the number of defects per board (10 m^2). x is Poisson distributed.

$$P(x) = \frac{\lambda^x e^{-\lambda}}{x!} \quad x = 0, 1, 2, \ldots$$

$$\begin{aligned} \lambda &= \frac{\text{average number of defects}}{\text{board}} \\ &= \left(\frac{10 \text{ m}^2}{1 \text{ board}}\right)\left(0.015 \frac{\text{defect}}{\text{m}^2}\right) \\ &= 0.15 \text{ defects/board} \end{aligned}$$

$$\begin{aligned} P\begin{pmatrix}\text{board} \\ \text{defective}\end{pmatrix} &= P(x > 1) = 1 - P(x \leq 1) \\ &= 1 - P(x = 0) - P(x = 1) \\ &= 1 - \frac{(0.15)^0 e^{-0.15}}{0!} - \frac{(0.15)^1 e^{-0.15}}{1!} \\ &= 1 - 0.8607 - 0.1291 \\ &= 0.0102 \\ P\begin{pmatrix}\text{not} \\ \text{defective}\end{pmatrix} &= 1 - 0.0102 = 0.9898 \end{aligned}$$

Let y = number of defects in a box. y is binomial.

$$\begin{aligned} P(y) &= \binom{50}{y} (0.0102)^y (0.9898)^{50-y} \\ & \quad y = 0, 1, 2, \ldots, 50 \\ P(y < 2) &= P(y \leq 1) = P(y = 0) + P(y = 1) \\ &= \binom{50}{0} (0.0102)^0 (0.9898)^{50} \\ & \quad + \binom{50}{1} (0.0102)^1 (0.9898)^{49} \\ &= 0.5989 + 0.3086 \\ &= 0.9075 \quad (0.91) \end{aligned}$$

Answer is D.

11. A product consists of two parts that are placed end to end. Assume that the dimensions of the length of the parts are approximately normally distributed with the mean and standard deviation shown as follows.

	mean length (in)	standard deviation (in)
part A:	2.65	0.12
part B:	1.45	0.38

The probability that the combined length is greater than 4.35 in is approximately

(A) 0.20
(B) 0.26
(C) 0.55
(D) 0.90

Solution:

Let μ_{A} = mean length of part A, and let μ_{B} = mean length of part B.

$$\begin{aligned} \sigma_{\text{A}}^2 &= \text{variance of length of part A} \\ \sigma_{\text{B}}^2 &= \text{variance of length of part B} \\ t &= \text{total length} \\ x_{\text{A}} &= \text{length of part A} \\ x_{\text{B}} &= \text{length of part B} \\ t &= x_{\text{A}} + x_{\text{B}} \end{aligned}$$

$f(t)$ is normally distributed.

$$E(t) = E(x_\text{A}) + E(x_\text{B}) = 2.65 + 1.45 = 4.1$$

The variance of t is

$$\sigma_\text{A}^2 + \sigma_\text{B}^2 = (0.12)^2 + (0.38)^2 = 0.1588$$

For $P(t > 4.35)$, the variance normal variant is

$$z = \frac{4.35 - 4.10}{\sqrt{0.1588}} = 0.63$$

$$P(z > 0.63) = 0.26$$

Answer is B.

12. Two different machines (A and B) are used to fill 20 oz containers of soda. The following data was collected. (All values are in oz.)

	bottles sampled	sample mean	sample variance
machine A:	16	20.25	0.18
machine B:	10	20.15	0.34

The endpoints of the 90% confidence interval estimate of the ratio of the population of variance A to variance B are

(A) 0.729, 5.682
(B) 0.176, 1.371
(C) 0.126, 3.821
(D) 0.177, 4.862

Solution:

σ_A^2 = variance for machine A
σ_B^2 = variance for machine B
s_A^2 = sample variance for machine A
s_B^2 = sample variance for machine B

Use the F distribution to estimate σ_B^2.

$$\frac{1}{F_{0.05,9,15}} \le \frac{\sigma_\text{A}^2 s_\text{B}^2}{\sigma_\text{B}^2 s_\text{A}^2} \le F_{0.05,9,15}$$

$$\frac{s_\text{A}^2}{(F_{0.05,9,15}) s_\text{B}^2} \le \frac{\sigma_\text{A}^2}{\sigma_\text{B}^2} \le \frac{s_\text{A}^2 (F_{0.05,9,15})}{s_\text{B}^2}$$

$$\frac{0.18}{(0.34)(3.01)} \le \frac{\sigma_\text{A}^2}{\sigma_\text{B}^2} \le \left(\frac{0.18}{0.34}\right)(2.59)$$

$$0.176 \le \frac{\sigma_\text{A}^2}{\sigma_\text{B}^2} \le 1.371$$

Answer is B.

FACILITY DESIGN AND LOCATION

13. Three products are produced on four machines (A, B, C, D) where each machine requires a 3.5 m × 3.5 m area. The routing for each product is given in the following table. Which block layout minimizes the overall flow × distance (assume centroid-to-centroid distances)?

	route				loads/day
product 1:	A	B	D	C	30
product 2:	B	C	D	A	20
product 3:	C	A	B	D	40

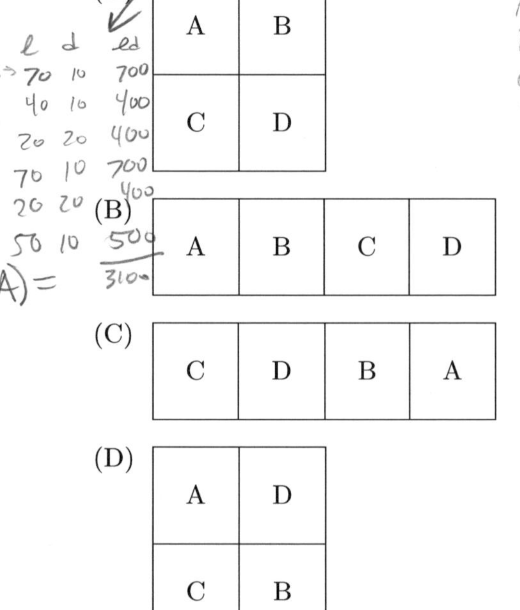

Solution:

The following is the upper triangular flow matrix (i.e., flow C-D of 50 = flow from C to D [product 2 = 20] and flow from D to C [product 1 = 30]).

from/to	A	B	C	D
A	–	30 + 40 = 70	40	20
B	–	–	20	30 + 40 = 70
C	–	–	–	30 + 20 = 50

Calculate the flow × distance score for each solution.

$$\begin{aligned} f \times d \text{ of solution A} &= (70 \times 10) + (40 \times 10) \\ &\quad + (20 \times 20) + (20 \times 20) \\ &\quad + (70 \times 10) + (50 \times 10) \\ &= 3100 \end{aligned}$$

$$\begin{aligned} f \times d \text{ of solution B} &= (70 \times 10) + (40 \times 20) \\ &\quad + (20 \times 30) + (20 \times 10) \\ &\quad + (70 \times 20) + (50 \times 10) \\ &= 4700 \end{aligned}$$

$$\begin{aligned} f \times d \text{ of solution C} &= (70 \times 10) + (40 \times 30) \\ &\quad + (20 \times 20) + (20 \times 20) \\ &\quad + (70 \times 10) + (50 \times 10) \\ &= 3900 \end{aligned}$$

$$\begin{aligned} f \times d \text{ of solution D} &= (70 \times 20) + (40 \times 10) \\ &\quad + (20 \times 10) + (20 \times 10) \\ &\quad + (70 \times 10) + (50 \times 20) \\ &= 3900 \end{aligned}$$

Answer is A.

14. A test station is to be incorporated into a fabrication department. The station will receive five loads per day from manufacturing center MC1 at a location given by (x, y) coordinates $(10, 8)$, four from MC2 at $(2, 4)$, three from MC3 at $(4, 6)$, and two from MC4 at $(6, 10)$. Where should the station be located to minimize total distance traveled?

(A) at $(6, 6)$
(B) at $(4, 6)$
(C) at $(6, 4)$
(D) at $(4, 8)$

Solution:

The optimal value of the test station can be found by solving for each of the two coordinates (x and y) separately using two mathematical properties of the optimal solution.

(1) The x- and y-coordinate of the new facility will be one of the x- and y-coordinates of the existing facilities.

(2) The optimal x and y will be located so that no more than half of the total weight is to the left of x and no more than half is to the right of x.

The application of these two properties gives the following.

$$\tfrac{1}{2} \text{ sum (or total weight)} = \frac{5+4+3+2}{2} = 7$$

	x-coordinate	weight	sum weights
MC2:	2	4	3
MC3:	4	3	7 = 7
MC4:	6	2	9
MC1:	10	5	14

The x-coordinate is 4.

	y-coordinate	weight	sum weights
MC2:	4	3	3
MC3:	6	4	7 = 7
MC1:	8	5	12
MC4:	10	2	14

The y-coordinate is 6.

Answer is B.

15. Given the following product demands, how many storage spaces would be required for all four products if a dedicated storage policy was used versus a random storage policy?

period	product 1	product 2	product 3	product 4
1	30	25	5	50
2	40	10	20	30
3	50	15	15	40
4	30	20	30	20
5	20	30	25	25

(A) dedicated = 100, random = 50
(B) dedicated = 120, random = 160
(C) dedicated = 160, random = 100
(D) dedicated = 160, random = 120

Solution:

period	product 1	product 2	product 3	product 4	total
1	30	25	5	50	110
2	40	10	20	30	100
3	50	15	15	40	120
4	30	20	30	20	100
5	20	30	25	25	100
maximum:	50	30	30	50	120

Dedicated storage space is equal to the sum of the maximum demands for each product.

$$50 + 30 + 30 + 50 = 160$$

Random storage space requirements are equal to the maximum aggregate demand over all time periods.

$$\text{maximum of } (110, 100, 120, 100, 100) = 120$$

Answer is D.

INDUSTRIAL COST ANALYSIS

16. A production company incurs fixed costs of \$1,250,000 per year. The variable cost of production is \$10 per unit. All units produced are sold for \$30. A maximum of 130,000 units can be manufactured each year. The break-even production quantity for the company is

(A) 47,500 units/yr
(B) 50,000 units/yr
(C) 62,500 units/yr
(D) 125,000 units/yr

Solution:

The break-even production quantity yields a total profit equal to zero.

$$\text{profit} = 0 = (p - c_v)D - C_F$$

$$D = \frac{C_F}{p - c_v}$$

p = selling price = \$30/unit

c_v = variable production cost = \$10/unit

C_F = fixed production costs = \$1,250,000/yr

D = production quantity (units/yr)

$$= \frac{\frac{\$1{,}250{,}000}{\text{yr}}}{\frac{\$30}{\text{unit}} - \frac{\$10}{\text{unit}}} = 62{,}500 \text{ units/yr}$$

Answer is C.

17. An industrial engineer is in charge of determining the standard labor cost of a new product. Experience has shown that a 90% learning curve applies to this type of product. The standard time used to compute the standard cost is taken to be the time to produce the 100th unit. The time to produce the first unit is 10 hr. Direct labor is paid wages of \$10/hr, and fringe benefits amount to 30% of the direct wage. Based on this information, the standard labor cost per unit for the new product should be

(A) \$64.56
(B) \$82.03
(C) \$103.26
(D) \$117.00

Solution:

$$\text{standard labor cost} = (\text{standard labor time}) \times (\text{labor cost per time unit})$$

$$\text{standard labor time} = \text{time to produce 100th unit}$$

$$Z_u = K u^{\frac{\log s}{\log 2}}$$

u = output unit number = 100

K = time to produce the first output unit = 10 hr

s = learning curve slope parameter expressed as a decimal = 0.90

Z_u = time to produce output unit number u

$$Z_{100} = (10 \text{ hr})(100)^{\frac{\log 0.9}{\log 2}} = (10 \text{ hr})(100)^{-0.152} = 4.966 \text{ hr}$$

wages = \$10/hr

fringe benefits = 30% of wages = (0.3)(wages)

$$\frac{\text{labor cost}}{\text{hr}} = \text{wages} + \text{fringe benefits} = \frac{\$10}{\text{hr}} + (0.3)\left(\frac{\$10}{\text{hr}}\right) = \$13/\text{hr}$$

$$\text{standard labor cost} = (4.966 \text{ hr})\left(\frac{\$13}{\text{hr}}\right) = \$64.56$$

Answer is A.

18. Consider the following cost data related to the production of a new product.

direct labor: 0.25 hr/unit at \$17/hr
direct material: \$300 per 50 units
overhead: 150% of total direct costs

If the company desires to make a profit equal to 12% of the total manufacturing cost, the selling price should be set to

(A) \$17.22/unit
(B) \$25.24/unit
(C) \$26.86/unit
(D) \$28.70/unit

Solution:

$$\frac{\text{selling price}}{\text{unit}} = \frac{\text{total manufacturing cost}}{\text{unit}} + \frac{\text{profit}}{\text{unit}}$$

$$\frac{\text{total manufacturing cost}}{\text{unit}} = \frac{\text{direct labor cost}}{\text{unit}} + \frac{\text{direct material cost}}{\text{unit}} + \frac{\text{overhead cost}}{\text{unit}}$$

$$\frac{\text{direct labor cost}}{\text{unit}} = \left(\frac{0.25 \text{ hr}}{\text{unit}}\right)\left(\frac{\$17}{\text{hr}}\right) = \$4.25/\text{unit}$$

$$\frac{\text{direct material cost}}{\text{unit}} = \frac{\$300}{50 \text{ units}} = \$6/\text{unit}$$

$$\begin{aligned}\frac{\text{overhead cost}}{\text{unit}} &= (150\%)\left(\frac{\text{direct labor cost}}{\text{unit}} + \frac{\text{direct material cost}}{\text{unit}}\right)\\ &= (1.5)\left(\frac{\$4.25}{\text{unit}} + \frac{\$6}{\text{unit}}\right)\\ &= \$15.375/\text{unit}\end{aligned}$$

$$\begin{aligned}\frac{\text{total manufacturing cost}}{\text{unit}} &= \frac{\$4.25}{\text{unit}} + \frac{\$6}{\text{unit}} + \frac{\$15.375}{\text{unit}}\\ &= \$25.625/\text{unit}\end{aligned}$$

$$\begin{aligned}\frac{\text{profit}}{\text{unit}} &= (12\%)\left(\frac{\text{total manufacturing cost}}{\text{unit}}\right)\\ &= (0.12)\left(\frac{\$25.625}{\text{unit}}\right)\\ &= \$3.075/\text{unit}\end{aligned}$$

$$\begin{aligned}\frac{\text{selling price}}{\text{unit}} &= \frac{\$25.625}{\text{unit}} + \frac{\$3.075}{\text{unit}}\\ &= \$28.70/\text{unit}\end{aligned}$$

Answer is D.

INDUSTRIAL ERGONOMICS

19. An alarm produces a 3000 Hz sound and a sound level of 70 dB, measured on the A scale and based upon a formula for a person with normal hearing. The area in the plant where this alarm is to be installed has both male and female employees ages 25 through 65. To ensure that all employees can hear this alarm, the sound level should be increased to approximately how many decibels measured on the A scale?

(A) The sound level should be increased to at least 75 dBA.
(B) The sound level should be increased to at least 90 dBA.
(C) The sound level should be increased to at least 100 dBA.
(D) The sound level should be increased to at least 120 dBA.

Solution:

The answer must address the average shifts in age of the threshold of hearing for pure tones of persons with normal hearing. Using the graphs in the *FE Reference Handbook* and designing for the worst condition at age 65,

shift for men at age 65: 30 dBA
shift for women at age 65: 22 dBA

Add 30 dBA to the current sound level of the alarm.

$$70 \text{ dBA} + 30 \text{ dBA} = 100 \text{ dBA}$$

Answer is C.

20. An adjustable seat is being designed to accommodate both males and females. Assume shoe height is 3 cm and the design needs to adjust to accommodate 5% to 95% of both males and females. The adjustment range in centimeters should be about

(A) 39–52 cm
(B) 35–55 cm
(C) 39–42 cm
(D) 37–51 cm

Solution:

Use the ergonomic table in the *FE Reference Handbook*.

Seating height plus 3 cm for shoes is as follows.

	percentiles	
	5th	95th
female:	38.5	47.3
male:	42.2	51.8

Use 38.5–51.8 cm (39–52 cm).

Answer is A.

21. A company has selected a standard chair height of 48 cm and a standard table height of 74 cm. Any tables under consideration must have a 3 cm thick top. What would be the maximum thickness of the support for the table top if 95% of all employees must be able to put their legs under the table? Assume shoe thickness is 3 cm.

(A) 3.5 cm
(B) 5.3 cm
(C) 6.8 cm
(D) 10.2 cm

Solution:

Look at the thigh thickness 95th percentile.

$$3 \text{ cm} + h + T + 48 \text{ cm} = 74 \text{ cm}$$

$$\begin{aligned} 48 \text{ cm} &= \text{height of chair} \\ 74 \text{ cm} &= \text{height of table} \\ 3 \text{ cm} &= \text{thickness of table} \\ h &= \text{thickness of support in cm} \\ T &= \text{thigh clearance height in cm} \end{aligned}$$

From the ergonomics table in the *FE Reference Handbook*, the thigh thickness (T) 95th percentile is

$$\begin{aligned} \text{men: } & 17.7 \text{ cm} \\ \text{women: } & 17.5 \text{ cm} \end{aligned}$$

$$T = 17.7 \text{ cm} \left(\begin{array}{c} \text{the larger of the values} \\ \text{for men and women} \end{array} \right)$$

$$3 \text{ cm} + h + 17.7 \text{ cm} + 48 \text{ cm} = 74 \text{ cm}$$

$$h = 5.3 \text{ cm}$$

Answer is B.

INDUSTRIAL MANAGEMENT

22. Motivating workers to perform efficiently and effectively has long been a problem facing industrial managers. A significant 1940's research project introduced what is called the "hierarchy-of-needs theory." As illustrated in the following figure, this theory recognized that people are motivated by five distinct types of needs: physiological, safety, love, esteem, and self-actualization.

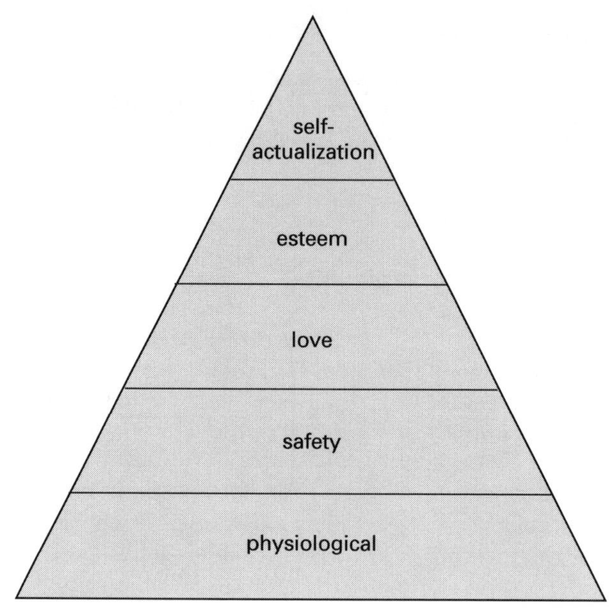

The researcher whose name is now synonymous with the hierarchy-of-needs theory is

(A) Frederick W. Taylor
(B) Frank B. Gilbreth
(C) Peter F. Drucker
(D) Abraham H. Maslow

Solution:

Abraham H. Maslow first published his hierarchy-of-needs theory in the journal *Psychological Review* in 1943.

Answer is D.

23. A recent industrial engineering graduate was hired by a large aircraft manufacturing company. The engineer works in the systems engineering department where the department manager ensures that the engineer has the necessary systems tools and contacts with other systems experts to perform her job. Shortly after being hired, the engineer is assigned to a large project involving the design and manufacture of a new commercial cargo plane. The cargo plane project manager employs the engineer's skills in combination with the skills from other disciplines to tackle the project. The engineer has, in essence, two bosses: the department manager and the project manager.

The organizational structure the engineer is experiencing at the aircraft company could best be described as a

(A) functional organization
(B) line-staff organization
(C) matrix organization
(D) network organization

Solution:

A matrix organizational structure applies dual chains of command. Functional departmentalization is used to gain economies from specialization. Overlaying the functional departments is a set of managers responsible for specific projects or products.

Answer is C.

24. Business process reengineering (BPR) has been defined as the fundamental analysis and radical redesign of strategic, value-added business processes and the systems, policies, and organizational structures that support them, resulting in a dramatic improvement in critical measures of business performance. In fact, some have labeled BPR as "the modern definition of Industrial Engineering." Yet, recent studies have shown that as few as one-third of BPR efforts are successful.

Experts in the field of BPR have identified ten primary reasons why BPR efforts fail. Which of the following is not a primary reason for failures in BPR efforts?

(A) BPR work can conflict with team members' "real jobs" or with other improvement programs.
(B) BPR team members lack the engineering skills to design and implement large-scale system changes.
(C) BPR efforts are attempted without defined methodologies.
(D) BPR teams consist of only selected representatives rather than people from all levels of all affected departments.

Solution:

Unlike continuous improvement programs, BPR involves radical and fundamental change. While it is essential to consider the human dimension of such radical and fundamental change, it is generally thought that involving representatives from all levels of all affected departments is a mistake. To use a medical analogy, continuous improvement is like vitamins and exercise while BPR is like radical surgery. Such surgery is best performed by a small, skilled team rather than an all-encompassing committee.

Answer is D.

INFORMATION SYSTEMS DESIGN

25. Management has approved a 10 megabytes/sec, Ethernet-based local area network for a manufacturing facility. The engineer has been asked to assist in defining the specifications for the network. Which set of specifications are appropriate?

(A) high-speed, dedicated, packet switching private branch exchange
(B) serial modems with even parity and serial line interface protocol access
(C) peer-to-peer, bus topology, carrier-sensing multiple access/collision detect access
(D) master-slave, ring topology, token-passing access

Solution:

The standard Ethernet protocol specifies a peer-to-peer local area network configuration with a bus topology and a carrier-sensing multiple access/collision detect access method. Choice (D) is also a local area network configuration, but a token-passing access protocol and a ring topology are not considered standard Ethernet specifications. Choices (A) and (B) are not local area network specifications and are considered analog phone line-based communication interfaces with a much slower data transmission rate.

Answer is C.

26. Assuming that sound has a frequency of 4000 Hz and a pulse code modulation samples sound at double the frequency, what is the data communications requirement (in bits/sec) for a 16-bit sound card?

(A) 14,400 bits/sec
(B) 28,800 bits/sec
(C) 128,000 bits/sec
(D) 256,000 bits/sec

Solution:

$$\begin{aligned}&\big((2)(\text{frequency})\big)(\text{no. of bits per sample})\\&\quad= \left((2)\left(4000\ \frac{1}{\text{sec}}\right)\right)\left(16\ \frac{\text{bits}}{\text{sample}}\right)\\&\quad= 128{,}000\ \text{bits/sec}\end{aligned}$$

A pulse code modulation sampling of twice the frequency for sound using a 16-bit sound card results in a data rate of 128 Kbps.

Answer is C.

27. In an entity-relationship diagram for a relational database, the relationship between a product record and the part family record it belongs to will be most likely represented as

(A) many to one (n:1)
(B) one to many (1:n)
(C) one to one (1:1)
(D) many to many (n:m)

Solution:

In group technology, a product can belong to only one part family (:1), but the part family can have many products as members of the family (n:). Thus, the relationship between the product record and the part family record it belongs to is many to one (n:1).

Answer is A.

MANUFACTURING PROCESSES

28. A single-point tool is used on an engine lathe to make a single pass on a 2 in diameter (before the cut), 24 in long part. The depth of cut is set to 0.125 in at a spindle speed of 400 rpm with a feed of 0.012 in per revolution.

The metal removal rate is most nearly

(A) 0.042 in^3/min
(B) 3.5 in^3/min
(C) 15 in^3/min
(D) 58 in^3/min

Solution:

$$
\begin{aligned}
A &= \text{cross-sectional area} \\
V &= \text{volume} \\
t &= \text{time} \\
f &= \text{feed} \\
s &= \text{spindle speed} \\
r_o &= \text{outer radius} \\
r_i &= \text{inner radius} \\
\text{MRR} &= \text{metal removal rate} \\
&= \frac{V}{t} = Afs = \pi(r_o^2 - r_i^2)fs \\
&= \pi\left[(1.000\ \text{in})^2 - (0.875\ \text{in})^2\right] \\
&\quad \times \left(0.012\ \frac{\text{in}}{\text{rev}}\right)\left(400\ \frac{\text{rev}}{\text{min}}\right) \\
&= 3.534\ \text{in}^3/\text{min} \quad (3.5\ \text{in}^3/\text{min})
\end{aligned}
$$

Answer is B.

29. The following diagram shows the top, side, and front views of a single-point cutting tool.

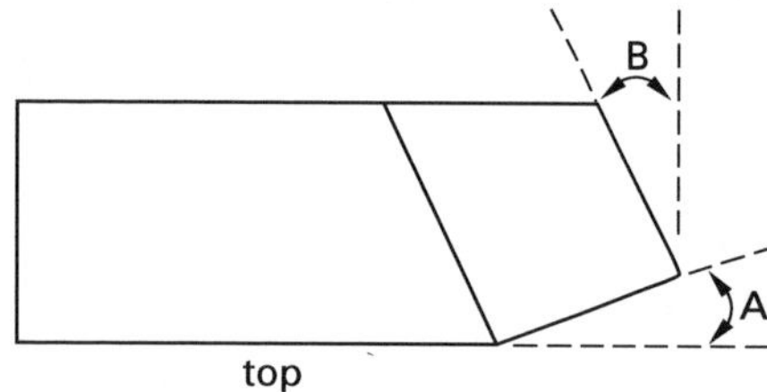

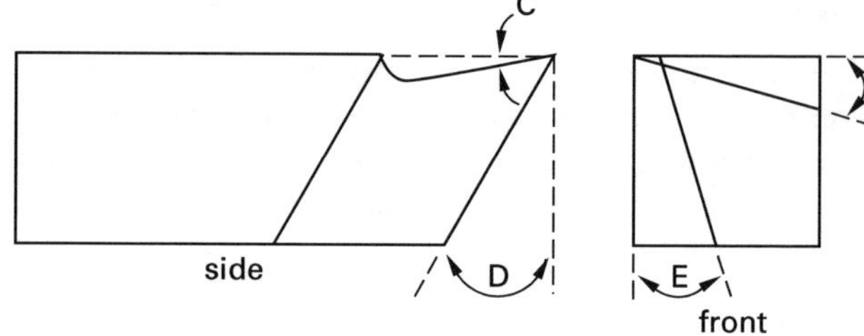

What is the angle labeled E called?

(A) the end cutting edge angle
(B) the end relief angle
(C) the side relief angle
(D) the side rake angle

Solution:

The side relief angle provides relief for the side cutting edge to keep the tool from rubbing against the cut surface.

Answer is C.

30. The following figures show a front view and a top view of the sleeve and thimble from a standard 1 inch micrometer.

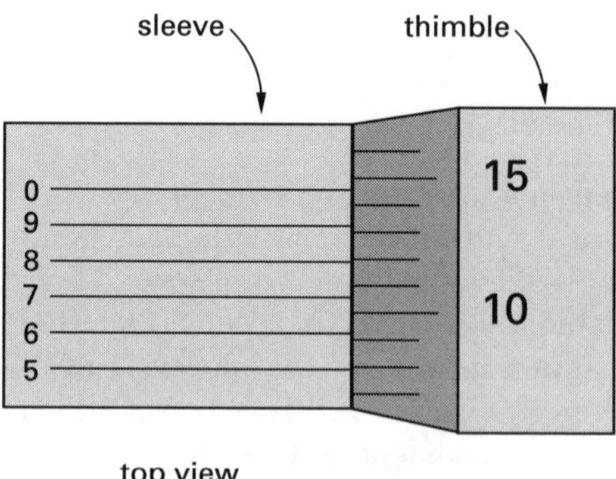

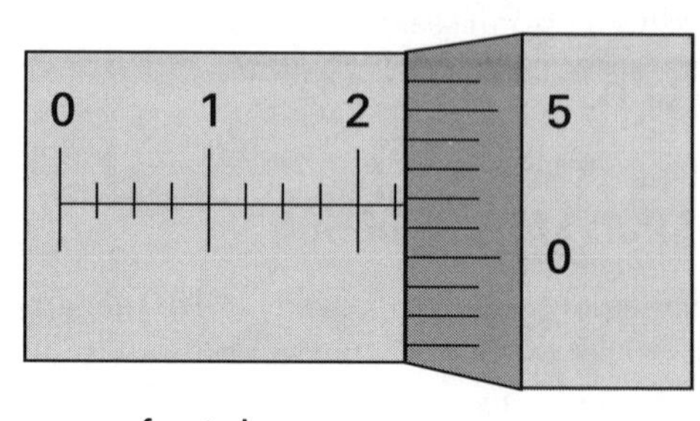

To the nearest $^1/_{10,000}$ of an inch, what is the micrometer reading?

(A) 0.2268 in
(B) 0.2520 in
(C) 0.3528 in
(D) 2.2518 in

Solution:

The front view shows 0.225 in + 0.001 in and the top view shows 0.0008 in, for a total of 0.2268 in.

Answer is A.

MANUFACTURING SYSTEMS DESIGN

31. The design for a new part arrives at a company's manufacturing engineering department for process planning. The design is available both as a computer-aided design file in product data exchange standard/initial graphics exchange specification format and as a blueprint hardcopy. The manufacturing engineering department has a computer-aided variant process planning system in place. Which set of steps should be used in preparing a process plan for this new part?

(A) Working from the computer-aided design file, the engineer should generate a process plan, and then assign it to a part family.
(B) The engineer should assign a group technology code to the part, look up the part family, and then modify the standard process plan.
(C) The engineer should assign the part to a part family and then generate the process plan.
(D) The engineer should create a part family matrix, look up the group technology code, and then modify the standard process plan.

Solution:

In a variant process planning system, the first step is to assign a group technology code to the new part. This code will then enable a lookup for a matching part family. If a matching part family is found, its standard plan can be modified specific to the features of the new part. Choices (A) and (C) refer to a generative process planning system that is of an altogether different nature. Choice (D) is not correct either, because a part family matrix should already be available in the existing variant process planning system and is not necessary for every new part.

Answer is B.

32. Suppose it is desired to scale the following object to three times its size with respect to point A.

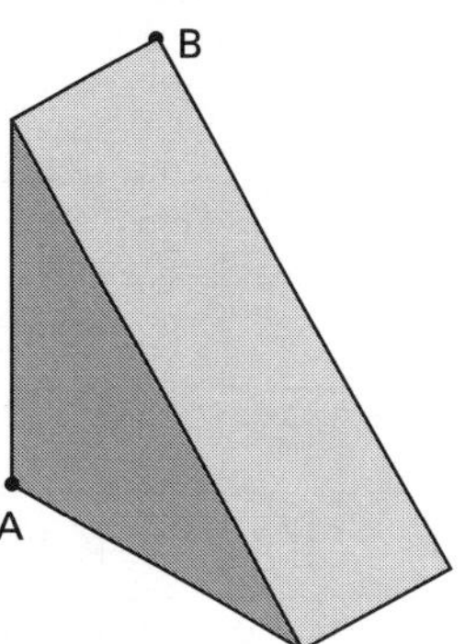

If point A has coordinates A(2, 2, 2) and point B has coordinates B(3, 3, 3), what would be the new coordinates for points A and B after scaling?

(A) A(2, 2, 2), B(5, 5, 5)
(B) A(6, 6, 6), B(9, 9, 9)
(C) A(5, 5, 5), B(3, 3, 3)
(D) A(2, 2, 2), B(9, 9, 9)

Solution:

With the scaling being done with respect to point A, point A does not change coordinates. The new coordinates of point B are calculated by performing the homogeneous matrix multiplication $\mathbf{T}^{-1} \times \mathbf{S} \times \mathbf{T}' \times \mathbf{B}$, where $\mathbf{B}$ is the vector form of point B, $\mathbf{T}'$ is the translation matrix that moves the reference point A to the origin, $\mathbf{S}$ is the scaling matrix, and $\mathbf{T}^{-1}$ is the translation matrix that returns the reference point A to its original position. The new coordinates are

$$\mathbf{T}^{-1} \times \mathbf{S} \times \mathbf{T}' \times \mathbf{B}$$

$$\begin{bmatrix} B_x \\ B_y \\ B_z \\ 1 \end{bmatrix} = \begin{bmatrix} 1 & 0 & 0 & 2 \\ 0 & 1 & 0 & 2 \\ 0 & 0 & 1 & 2 \\ 0 & 0 & 0 & 1 \end{bmatrix} \begin{bmatrix} 3 & 0 & 0 & 0 \\ 0 & 3 & 0 & 0 \\ 0 & 0 & 3 & 0 \\ 0 & 0 & 0 & 1 \end{bmatrix} \begin{bmatrix} 1 & 0 & 0 & -2 \\ 0 & 1 & 0 & -2 \\ 0 & 0 & 1 & -2 \\ 0 & 0 & 0 & 1 \end{bmatrix} \begin{bmatrix} 3 \\ 3 \\ 3 \\ 1 \end{bmatrix} = \begin{bmatrix} 5 \\ 5 \\ 5 \\ 1 \end{bmatrix}$$

Answer is A.

33. Using the production flow analysis method (or any other equivalent clustering procedure), determine the appropriate machine cell and part family grouping for the following machine/part matrix.

	C1	C2	C3	C4	C5
M1		1		1	
M2	1	1	1	1	1
M3		1		1	
M4	1		1		1
M5	1				1

(A) family/cell A (C2, C4/ M2, M4, M5); family/cell B (C1, C3, C5/ M1, M3)
(B) family/cell A (C2, C4/ M1, M3); family/cell B (C1, C3, C5/ M2, M4, M5)
(C) family/cell A (C1, C5/ M2, M4, M5); family/cell B (C2, C3, C4/ M1, M2, M3, M4)
(D) family/cell A (C1, C3, C5/ M2, M4, M5); family/cell B (C2, C3, C4/ M1, M2, M3)

Solution:

Using Burbridge's production flow analysis method that assigns binary weights to the rows and columns iteratively until ordering is achieved (or using any other equivalent clustering method such as similarity coefficient weighting) will ultimately lead to a grouping of machines and parts as given in choice (B).

	C1	C2	C3	C4	C5	
M1		1		1		2^0
M2	1	1	1	1	1	2^1
M3		1		1		2^2
M4	1		1		1	2^3
M5	1				1	2^4
	26	7	10	7	26	

	C2	C4	C3	C1	C5	
M1	1	1				3
M2	1	1	1	1	1	31
M3	1	1				3
M4			1	1	1	28
M5				1	1	24
	2^0	2^1	2^2	2^3	2^4	

	C2	C4	C3	C1	C5	
M1	1	1				2^0
M3	1	1				2^1
M5				1	1	2^2
M4			1	1	1	2^3
M2	1	1	1	1	1	2^4
	19	19	24	28	28	

Answer is B.

MATERIAL HANDLING SYSTEMS DESIGN

34. It is most appropriate and economical to use a conveyor-based material handling system when

(A) the material flow rate is medium to high, the distance is short to medium, and the product routing is fixed
(B) the material flow rate is low to medium, the distance varies from short to medium, and the product routing is variable but well defined
(C) the material flow rate is low, the distance is highly variable, and the product routing is highly variable
(D) the material flow rate is high, the distance is long, and the product routing is moderately variable

Solution:

Conveyors are most economical when used for high-flow rate moves over short distances. Since the cost to reconfigure a conveyor system is high, conveyors are most appropriate for fixed product routings.

Answer is A.

35. A recently hired industrial engineer submitted a material handling system design that contains the following processes: (1) transfer, where a gravity conveyor moves a pallet holding a 100 kg, 0.5 m × 0.5 m × 0.5 m box containing four 25 kg gear casings delivered from a long-term contract vendor from the loading dock to incoming inspection 20 m away, (2) incoming inspection, where the cardboard containers for the boxes are discarded and the casings are manually loaded onto an overhead power and free conveyer, and (3) transfer, where the casings are transported in a loop from workstations 1 to 2 to 3 to 4 to 5 to packaging. The system automatically moves parts to the appropriate workstation based on bar code readings. At the packaging station, ten finished 25 kg casings are reboxed and placed on a pallet to be moved by forklift into a waiting truck. Which of the following lists all of the material handling design principles that were violated?

(A) mechanization, unit load, and simplification
(B) flexibility, systems, and gravity
(C) flexibility, ergonomics, and ecology
(D) ergonomics, gravity, unit load, and systems

Solution:

Flexibility was violated because a conveyor system is inherently inflexible and hard to reconfigure if future casing designs require more complex routings. Ergonomics was violated because the loading of a 25 kg casing is beyond the lifting limits for repetitive work. Ecology was violated because the vendor has an established relationship with the company, and the IE should have worked with the vendor to design a reusable container that would minimize both the cost and waste from transporting the casings.

Answer is C.

36. A factory with four departments has to determine the type of material handling equipment to use. The departments have determined that either a forklift, a pallet jack, or a conveyor system would be technically feasible. However, the combined capital and operating cost ($ per meter traveled) and the size of the unit load depend on the type of material handling equipment selected. The following material flow routing, distance, and material handling equipment data are given. What is the total cost associated with each type of equipment, and which should be selected?

material handling type	units/load	$/m
conveyor	100	0.2
forklift	200	1
pallet jack	20	0.05

	routing				no. of units
product 1:	D1	D2	D3	D4	1000
product 2:	D1	D3	D2	D4	2000

distance (m)	D1	D2	D3	D4
D1	–	50	100	150
D2		–	50	100
D3			–	50
D4				–

(A) conveyor = $2000, forklift = $2200, pallet jack = $2500; select the conveyor
(B) conveyor = $1300, forklift = $3250, pallet jack = $1625; select the forklift
(C) conveyor = $900, forklift = $1550, pallet jack = $550; select the pallet jack
(D) conveyor = $1300, forklift = $3250, pallet jack = $1625; select the conveyor

Solution:

The total distance for product 1 is

$$50 \text{ m} + 50 \text{ m} + 50 \text{ m} = 150 \text{ m}$$

The total distance for product 2 is

$$100 \text{ m} + 50 \text{ m} + 100 \text{ m} = 250 \text{ m}$$

For the conveyor, the total cost is

$$\left(0.2 \ \frac{\$}{\text{m}}\right)\left[\begin{array}{l}\left(\dfrac{1000 \text{ units}}{100 \ \dfrac{\text{units}}{\text{load}}}\right)(150 \text{ m}) \\ + \left(\dfrac{2000 \text{ units}}{100 \ \dfrac{\text{units}}{\text{load}}}\right)(250 \text{ m})\end{array}\right] = \$1300$$

For the forklift, the total cost is

$$\left(1\ \frac{\$}{\text{m}}\right)\left[\left(\frac{1000\ \text{units}}{200\ \frac{\text{units}}{\text{load}}}\right)(150\ \text{m}) + \left(\frac{2000\ \text{units}}{200\ \frac{\text{units}}{\text{load}}}\right)(250\ \text{m})\right] = \$3250$$

For the pallet jack, the total cost is

$$\left(0.05\ \frac{\$}{\text{m}}\right)\left[\left(\frac{1000\ \text{units}}{20\ \frac{\text{units}}{\text{load}}}\right)(150\ \text{m}) + \left(\frac{2000\ \text{units}}{20\ \frac{\text{units}}{\text{load}}}\right)(250\ \text{m})\right] = \$1625$$

Answer is D.

MATHEMATICAL OPTIMIZATION AND MODELING

37. For the following problem, which of the following is the optimal solution for both the original problem and its dual? Let y represent the dual values.

$$\begin{aligned} \text{min}\ & 3x_1 + 12x_2 \\ \text{subject to}\ & 2x_1 + 4x_2 - 2x_3 \geq 6 \\ & -6x_1 - 3x_2 + 12x_3 \geq 6 \\ & x \geq 0 \end{aligned}$$

(A) $x_1 = 7$, $x_2 = 0$, $x_3 = 4$, and $y_1 = 3$, $y_2 = \frac{1}{2}$
(B) $x_1 = 0$, $x_2 = 2$, $x_3 = 0$, and $y_1 = \frac{1}{2}$, $y_2 = 3$
(C) $x_1 = 7$, $x_2 = 0$, $x_3 = 4$, and $y_1 = 2$, $y_2 = \frac{1}{2}$
(D) $x_1 = 1$, $x_2 = 2$, $x_3 = 2$, and $y_1 = 4$, $y_2 = \frac{1}{2}$

Solution:

To be a solution, the points must satisfy the constraints (i.e., they must be feasible) of their respective problem, and the objective functions of the original and dual problems must be equal. The dual objective function for this problem is given by the right-hand side of the original problem and is $6y_1 + 6y_2$. The constraints for the dual are given by the columns and the objective function coefficients of the original problem. The dual has the form

$$\begin{aligned} \text{max}\ & 6y_1 + 6y_2 \\ \text{subject to}\ & 2y_1 - 6y_2 \leq 3 \\ & 4y_1 - 3y2 \leq 12 \\ & -2y_1 + 12y_2 \leq 0 \\ & y_1, y_2 \geq 0 \end{aligned}$$

The solution given in choice (A) satisfies the constraints, and the two objective functions are equal. For the original problem constraints,

$$(2)(7) + (4)(0) - (2)(4) = 6$$
$$(-6)(7) - (3)(0) + (12)(4) = 6$$

For the dual problem constraints,

$$(2)(3) - (6)\left(\tfrac{1}{2}\right) = 3$$
$$(4)(3) - (3)\left(\tfrac{1}{2}\right) = 10.5 \leq 12$$
$$(-2)(3) + (12)\left(\tfrac{1}{2}\right) = 0$$

The objective functions are

$$(3)(7) + (12)(0) = 21 = (6)(3) + (6)\left(\tfrac{1}{2}\right)$$

Note that the solution given by choice (B) is not feasible for the original problem. The solution given by choice (C) does not yield objective function values that are equal and the solution given by choice (D) is not feasible for the dual problem.

Answer is A.

38. The optimal solution for the following problem is given by

$$\begin{aligned} \text{max}\ & 5x_1 + x_2 \\ \text{subject to}\ & x_1 + 2x_2 \leq 10 \\ & -3x_1 + 4x_2 \leq 10 \\ & 3x_1 + 2x_2 \leq 18 \\ & x_1, x_2 \geq 0 \end{aligned}$$

(A) $x_1 = 4$ and $x_2 = 3$
(B) $x_1 = 10$ and $x_2 = 0$
(C) $x_1 = 6$ and $x_2 = 0$
(D) $x_1 = 5$ and $x_2 = 3$

Solution:

This is a simple two-dimensional problem and can be solved graphically to find the optimal basic feasible solution that corresponds to a vertex of the feasible region. The optimal vertex is the last feasible point that the objective function contours will touch as the value of the objective function is increased.

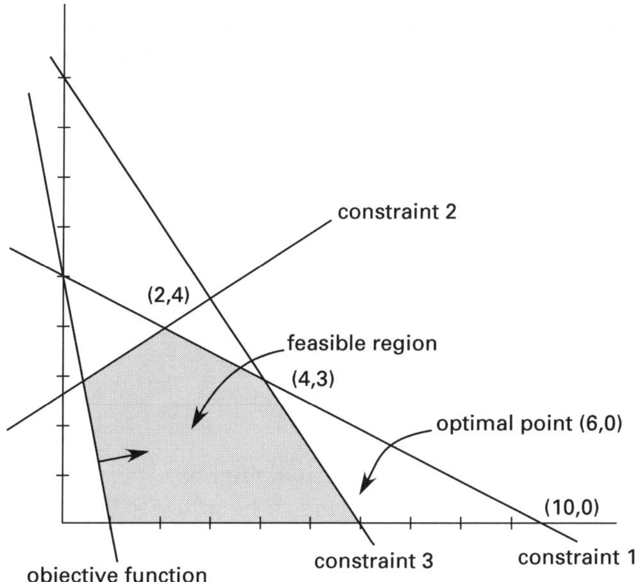

Answer is C.

39. Consider the following linear programming formulation.

$$\begin{aligned} \text{max}\ & 3x_1 + x_2 + 2x_3 \\ \text{subject to}\ & x_1 - x_2 + 2x_3 \leq 20 \\ & 2x_1 + x_2 - x_3 \leq 60 \\ & x \geq 0 \end{aligned}$$

The optimal tableau is given as follows, where x_4 and x_5 are the corresponding slack variables for constraints 1 and 2, respectively.

objective function coefficient		3	1	2	0	0	
c_B	x_B	x_1	x_2	x_3	x_4	x_5	b
2	x_3	3	0	1	1	1	30
1	x_2	5	1	0	1	2	40
reduced cost values	r_j	−8	–	–	−3	−4	

The most one would be willing to pay for additional units of constraint 2 is most nearly

(A) \$1
(B) \$2
(C) \$3
(D) \$4

Solution:

The most one would be willing to pay for additional units of a constraint is based upon the shadow price of that constraint. The shadow price comes from the dual value associated with the constraint. In the optimal tableau, the dual values can be obtained directly from the reduced costs associated with the slack variables. In this problem, x_5 is the slack variable associated with the second constraint. Therefore, the "value" of the second constraint is \$4. This means that for each increase of 1 unit of the second constraint, the objective function will increase by 4. Hence, as long as each additional unit of the second constraint costs less than \$4, it would be advantageous to purchase those units.

Answer is D.

PRODUCTION PLANNING AND SCHEDULING

40. Jobs go through a drilling station and then an automatic inspection station. Each station can only service one job at a time. Four jobs are currently awaiting processing. The time each job must spend at each station is shown as follows.

job	drilling time	inspection time
1	12	9
2	8	5
3	6	8
4	3	4

The job sequence that minimizes the makespan schedule is

(A) $\{4, 3, 2, 1\}$
(B) $\{4, 2, 1, 3\}$
(C) $\{4, 3, 1, 2\}$
(D) $\{4, 2, 3, 1\}$

Solution:

Johnson's rule results in the minimum makespan schedule for a two-machine flow shop. The following steps detail the application of Johnson's rule to this problem.

The shortest time is the drilling operation of job 4. Since this time is at the first workstation, job 4 is placed first in the sequence.

current sequence:

4			

The shortest time of remaining jobs is the inspection time of job 2. Since this time is at the second station, job 2 is placed last in the sequence.

current sequence:

4			2

The shortest time of remaining jobs is now the drilling operation of job 3. Since this time is at the first station, job 3 is placed behind job 4 at the beginning of the sequence.

current sequence:

4	3		2

Only job 1 remains. Job 1 is placed in the remaining slot in the sequence.

final sequence:

4	3	1	2

Answer is C.

41. The following table contains sales data for product X over the past twelve months.

month	units sold	month	units sold
1	100	7	120
2	93	8	108
3	102	9	83
4	78	10	96
5	85	11	115
6	98	12	110

Using a three-month moving average, the forecast for sales of product X during the next month is closest to

(A) 99 units
(B) 107 units
(C) 110 units
(D) 112 units

Solution:

The general equation for a moving average forecast is

$$F_t = \frac{1}{n}\sum_{i=1}^{n} D_{t-i}$$

F_t = forecast for period t
D_t = actual demand in period t
n = number of most recent observations to include in the forecast for the next period

For this problem, $t = 13$ and $n = 3$.

$$F_{13} = \tfrac{1}{3}\sum_{i=1}^{3} D_{13-i} = \left(\tfrac{1}{3}\right)(D_{12} + D_{11} + D_{10})$$
$$= \left(\tfrac{1}{3}\right)(110 \text{ units} + 115 \text{ units} + 96 \text{ units})$$
$$= 107 \text{ units}$$

Answer is B.

42. Five jobs are waiting to be processed at a machining center. The processing times and due dates (number of days remaining until job is due) for these jobs are as follows.

job number	processing time (days)	days remaining until job is due
1	2.00	4
2	1.50	8
3	4.00	5
4	3.75	6
5	0.75	2

What is the job sequence that minimizes the average time a job will spend waiting to be processed?

(A) $\{3, 5, 1, 4, 2\}$
(B) $\{5, 2, 1, 4, 3\}$
(C) $\{5, 1, 3, 4, 2\}$
(D) $\{3, 4, 1, 2, 5\}$

Solution:

Sequence jobs in nondecreasing order of processing time (Shortest Processing Time rule) to minimize the mean waiting time at the machining center.

job number	processing time
5	0.75
2	1.50
1	2.00
4	3.75
3	4.00

The sequence that minimizes the mean waiting time is $\{5, 2, 1, 4, 3\}$.

Answer is B.

PRODUCTIVITY MEASUREMENT AND MANAGEMENT

43. Which of the following lists contains only typical productivity tools used by company managers?

(A) supervisor training, job design, and incentive plans
(B) incentive plans, computer programming, and finite element analysis
(C) work standards, stock dividends, and statistical process control
(D) statistical process control, activity-based accounting, and discounted cash flow analysis

Solution:

Typical productivity tools used by company managers include training of both supervisors and employees, analyzing job design to make it most efficient, and instituting incentive plans to motivate employees.

Answer is A.

44. Which of the following lists contains only characteristics of successful incentive systems?

(A) ceilings on employee earnings, increased unit cost, and gain sharing with employer
(B) increased rate of production, fair standards, reduced unit cost, and increased employee earnings
(C) complicated pay formulas, good standards, and selective application to employees
(D) increased employee earnings, absence of union support, and unchanged rate of production

Solution:

A successful incentive system provides benefits to both management and the employee. Therefore, the system would be expected to increase production and associated productivity (which would reduce cost), be fair, be understandable to employees, and increase their earning potential.

Answer is B.

45. A company uses a piecework incentive system. The base rate is \$12.00 per hour for a 10 hr work day. For a particular assembly operation, the standard time per piece is 0.6 min. If a worker assembles 1150 pieces in one day, what would he be paid?

(A) \$115
(B) \$123
(C) \$138
(D) \$157

Solution:

The piecework rate is

$$\begin{aligned}(\text{base rate})\left(\frac{\text{standard time}}{\text{pieces}}\right) &= \left(\frac{\$12.00}{\text{hr}}\right)\left(\frac{0.6\text{ hr}}{60\text{ piece}}\right)\\ &= \$0.12/\text{piece}\end{aligned}$$

The pay is

$$\begin{aligned}&(\text{piecework rate})(\text{no. of pieces completed})\\ &\quad = (\$0.12)(1150)\\ &\quad = \$138\end{aligned}$$

Answer is C.

QUEUING THEORY AND MODELING

46. A national sports team uses an automated answering service for taking ticket orders. The service is always in one of two modes: answering the phone or off the phone. Calls come in at the rate of 10 per hour. The service can handle 15 calls per hour. Due to a high demand for tickets, the service runs 24 hours a day. What is the probability that the service will be answering a call 3 weeks from now?

(A) 0.33
(B) 0.40
(C) 0.50
(D) 0.60

Solution:

This is an example of a continuous Markov chain. Since the rates are per hour and the service runs 24 hours a day, 3 weeks from now represents a long-term horizon. Therefore, to find the probability that the service will be on the phone, you only need to find the steady-state probability for being on the phone. This is done through solving the system $\pi\Lambda = 0$ and $\Sigma\pi_i = 1$, where Λ is the transition rate matrix and π is the steady-state probability vector. The elements λ_{ij}, $i \neq j$, of the matrix Λ are the transition rates from state i to state j. The diagonal elements are given by $\lambda_{ii} = -(\Sigma_{j\neq i}\lambda_{ij})$. Here π_1 represents the probability of being in state 1 and π_2 represents the probability of being in state 2. For this problem, if state 1 is "off the phone" and state two is "on the phone," the transition rate matrix is

$$\Lambda = \begin{pmatrix} -10 & 10 \\ 15 & -15 \end{pmatrix}$$

The steady-state equations would be

$$(\pi_1, \pi_2)\begin{pmatrix} -10 & 10 \\ 15 & -15 \end{pmatrix} = (0, 0)$$

This yields the following set of equations.

$$\begin{aligned}-10\pi_1 + 15\pi_2 &= 0\\ 10\pi_1 - 15\pi_2 &= 0\end{aligned}$$

This system reduces to $-10\pi_1 + 15\pi_2 = 0$, which has the solution

$$\pi_1 = \tfrac{3}{2}\pi_2$$

By the normalizing equation ($\Sigma\pi_i = 1$);

$$\tfrac{3}{2}\pi_2 + \pi_2 = 1$$

This implies $\pi_2 = \frac{2}{5} = 0.4$.

$$\pi_1 = \tfrac{3}{5} = 0.6$$

Therefore, the probability of being on the phone (state 2) is 0.4.

Answer is B.

47. For the following transition diagram of a discrete Markov chain, identify whether the states are transient, absorbing, or recurrent.

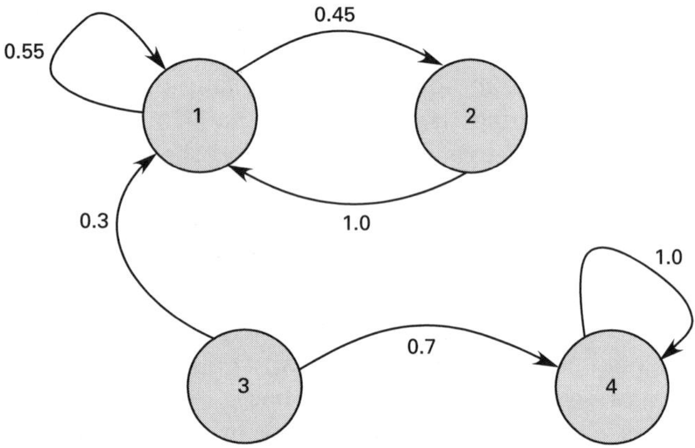

(A) All states are recurrent.
(B) State 1 is recurrent, states 2 and 3 are transient, and state 4 is absorbing.
(C) State 4 is absorbing and states 1, 2, and 3 are transient.
(D) States 1 and 2 are recurrent, state 3 is transient, and state 4 is absorbing.

Solution:

Recurrent means that a state communicates with itself. That is, once you leave the state you will eventually return to the state. *Transient* means that eventually you will never be able to return to the state. An *absorbing* state is recurrent and has a one-step transition probability of 1.0 of going from itself to itself. From these definitions it can be seen that state 4 is absorbing since its one-step transition probability is 1.0. State 3 is transient since a transition from state 3 either takes you to the absorbing state 4 or to a closed system of states 1 and 2 from which it is impossible to return to state 3. States 1 and 2 are recurrent since there is a possibility of returning to each of these states.

Answer is D.

48. The emergency room at a small local hospital has only one doctor on duty. The facility has three examination rooms and a waiting room that can hold seven people. Patients arrive according to a Poisson distribution with an arrival rate of ten people per hour. The doctor can examine and treat a patient according to an exponential distribution with a mean of 4 minutes. What is the average waiting time for a patient before they see the doctor?

(A) 4 min
(B) 4.12 min
(C) 7.25 min
(D) 20 min

Solution:

This is an M/M/1/10 model. The service and arrival times are exponentially distributed, there is one server, and the capacity of the system is 10 (9 people waiting and 1 being served). The average waiting time before a patient sees a doctor is given by W_q. By Little's law, $W_q = L_q/\lambda$.

For this system, the arrival rate, λ, is 10 per hour and the service rate, μ, is 15 per hour.

$$\rho = \frac{\lambda}{\mu} = \frac{\frac{10 \text{ people}}{1 \text{ hr}}}{\frac{15 \text{ people}}{1 \text{ hr}}} = \frac{2}{3}$$

To find W_q, find $L_q = L - (1 - P_0)$.

$$P_0 = \frac{1-\rho}{1-\rho^{10+1}} = \frac{\frac{1}{3}}{1-\left(\frac{2}{3}\right)^{11}} = 0.3372$$

$$L = \frac{\rho}{1-\rho} - \frac{(10+1)\rho^{10+1}}{1-\rho^{10+1}} = 2 - \frac{(11)\left(\frac{2}{3}\right)^{11}}{1-\left(\frac{2}{3}\right)^{11}} = 1.87135$$

$$L_q = 1.87135 - (1 - 0.3372) = 1.20855$$

$$W_q = \left(\frac{1.20855}{\frac{10 \text{ people}}{1 \text{ hr}}}\right)(60 \text{ min}) = 7.25 \text{ min}$$

Answer is C.

SIMULATION

49. Simulation experiments modeling a nonterminating system are often biased by the empty-and-idle startup conditions of the experiment which do not reflect the system's steady-state operating conditions.

Which of the following is not a viable method for dealing with initial condition bias when experimenting with a model of a nonterminating system?

(A) Discard statistics gathered during the startup phase.
(B) Perform a statistically significant number of replications of the experiment to overcome the startup bias.
(C) Run the experiment for a sufficiently long period such that the steady-state statistics effectively overwhelm those from the startup phase.
(D) Initialize the experiment states and conditions such that the startup conditions are representative of those of steady state.

Solution:

Replicating an experiment simply produces multiple data sets, each of which contain biases due to the initial conditions.

Answer is B.

50. The following data were gathered from two sets of experiments comparing entity times-in-system between two systems, A and B.

replication	system A	system B
1	9.19	8.87
2	19.18	31.52
3	12.30	14.14
4	13.04	14.11
5	17.79	16.72
6	11.49	27.78
7	21.61	34.05
8	10.50	22.96
9	13.27	10.98
10	8.50	11.66

A paired-t 95% confidence interval comparing time-in-system for system B with that for system A is most nearly

(A) $[0.80, 10.38]$
(B) $[-5.28, 5.28]$
(C) $[1.56, 9.62]$
(D) $[-10.13, 21.31]$

Solution:

First, calculate the individual replication differences, d_i.

replication	system A	system B	$d_i = \text{B} - \text{A}$
1	9.19	8.87	−0.32
2	19.18	31.52	12.34
3	12.30	14.14	1.84
4	13.04	14.11	1.07
5	17.79	16.72	−1.07
6	11.49	27.78	16.29
7	21.61	34.05	12.44
8	10.50	22.96	12.46
9	13.27	10.98	−2.29
10	8.50	11.66	3.16

The confidence interval is $[\bar{d} - h, \bar{d} + h]$.

$$\begin{aligned} d_i &= \text{differences between systems A and B} \\ n &= \text{number of replications} = 10 \\ t_{n-1,\,1-\alpha/2} &= \text{value of the student-}t\text{ distribution with} \\ &\quad\; n-1 \text{ degrees of freedom at level } \alpha \end{aligned}$$

$$\bar{d} = \frac{\sum\limits_{i=1}^{n} d_i}{n}$$

$$s(d) = \sqrt{\frac{\sum\limits_{i=1}^{n}(d_i - \bar{d})^2}{n-1}}$$

$$s(\bar{d}) = \frac{s(d)}{\sqrt{n}}$$

$$h = [(t_{n-1,\,1-\alpha/2})][s(\bar{d})]$$

For this problem, $n = 10$ and $\alpha = 0.05$.

$$\begin{aligned} \bar{d} &= \frac{\begin{array}{c}-0.32 + 12.34 + 1.84 + 1.07 - 1.07 \\ +\, 16.29 + 12.44 + 12.46 - 2.29 + 3.16\end{array}}{10} \\ &= 5.59 \end{aligned}$$

$$\begin{aligned} s(d) &= \sqrt{\frac{\begin{array}{c}(-0.32 - 5.59)^2 + (12.34 - 5.59)^2 \\ +\, (1.84 - 5.59)^2 + \cdots + (3.16 - 5.59)^2\end{array}}{10 - 1}} \\ &= \sqrt{\frac{435.95}{9}} \\ &= 6.96 \end{aligned}$$

$$s(\bar{d}) = \frac{6.96}{\sqrt{10}} = 2.12$$

$$t_{9,\,0.975} = 2.26 \quad \text{[from the student-}t\text{ table]}$$

$$h = (2.26)(2.12) = 4.79$$

The confidence interval is

$$[5.59 - 4.79, 5.59 + 4.79] = [0.80, 10.38]$$

Answer is A.

51. A data set has been gathered with 725 values measuring the number of arrivals per minute at a workstation. Calculations show that the data follow a Poisson distribution with a mean of 17 min.

What distribution and parameters, in minutes, should be used in a simulation model for the inter-arrival times at the workstation?

(A) normal with a mean of 17 and variance of $17/\sqrt{725}$
(B) Weibull with an alpha of 17 and a beta of 725
(C) Poisson with a mean of 17
(D) exponential with a mean of 1/17

Solution:

If the number of events that occur in a fixed time interval has a Poisson distribution with mean λ, then the time between events is exponentially distributed with a mean of $1/\lambda$.

Answer is D.

STATISTICAL QUALITY CONTROL

52. A company uses a single-attribute sampling plan on incoming critical materials.

$$\begin{aligned} N &= \text{lot size} = 50{,}000 \\ n &= \text{random sample size} = 3 \\ c &= \text{acceptance number} = 0 \end{aligned}$$

Three parts are drawn at random from a carload of 50,000 parts and are subjected to a stress yield test. If all three pieces meet the minimum yield strength, the entire lot is accepted. If one or more parts fail the test, the entire lot is returned to the supplier.

If the lot is submitted from a process whose population mean proportion defective rate is 10%, what is the probability of accepting this lot?

(A) 0%
(B) 10%
(C) 40%
(D) 75%

Solution:

Use the Poisson distribution to approximate the binomial probabilities.

$$\begin{aligned} P(x=0) &= \frac{e^{-nP}(nP)^x}{x!} = \frac{e^{-(3)(0.10)}\left((3)(0.10)\right)^0}{0!} \\ &= 0.7408 \quad (75\%) \end{aligned}$$

Answer is D.

53. The specification limits for an axle turned on a lathe had the following blueprint specifications for the diameter. (Note that it is possible to adjust the lathes, thus moving the population mean.)

$$285 \pm 6.0$$

A trial control chart on the process yielded the following limits for $\overline{X}$ and R based on 23 samples of size 5.

$$\begin{aligned} \text{UCL}_{\overline{X}} &= 294.16 \\ \text{LCL}_{\overline{X}} &= 283.76 \\ \text{UCL}_R &= 28.910 \\ \text{LCL}_R &= 0 \end{aligned}$$

All sample points fell within the trial control limits ($K = 23$).

$$\overline{\overline{X}} = \frac{\sum \overline{X}_i}{K} = 288.96$$

$$\overline{R} = \frac{\sum R}{K} = 8.960$$

Assuming that the population from which the data were drawn is approximately normally distributed, what is the population mean?

(A) 285.00
(B) 288.90
(C) 288.96
(D) 289.06

Solution:

$$\mu_x \approx \overline{\overline{X}} = 288.96$$

Answer is C.

54. Assuming permanent $\overline{X}$ and R control charts are needed for this process, the $\text{UCL}_{\overline{X}}$ and $\text{LCL}_{\overline{X}}$ limits should be set at most nearly

(A) 290.17; 279.83
(B) 291.00; 279.00
(C) 292.15; 281.05
(D) 294.16; 283.76

Solution:

The lathe cutting diameter can be adjusted, thus adjusting the population mean to a desired setting, $\mu_0 = 285.0$, the specified diameter.

$$\sigma_X \approx \frac{\overline{R}}{d_{2,n=5}} = \frac{8.960}{2.326} = 3.852$$

The permanent control chart limits for $\overline{X}$ would be $\mu_0 \pm 3(\sigma_{\overline{X}}/\sqrt{n})$ where $\sigma_X = 3.852$ and $n = 5$.

$$\text{UCL}_{\overline{X}} = 285.0 + (3)\left(\frac{3.852}{\sqrt{5}}\right) = 290.17$$

$$\text{LCL}_{\overline{X}} = 285.0 - (3)\left(\frac{3.852}{\sqrt{5}}\right) = 279.83$$

Answer is A.

TOTAL QUALITY MANAGEMENT

55. The ISO 9000 is best described as

(A) a top-to-bottom quality management system
(B) a system limited to foreign exporters/importers
(C) an easily implemented quality management system
(D) a quality management system applicable only to firms of 500 or more employees

Solution:

"Going through the ISO 9000 registration steps will give an organization a good start on implementing total quality." (Goetoch, D.L., and Davis, S.B., *Introduction to Total Quality*, Prentice-Hall, 2nd edition, 1997 (pp. 646-7).)

Answer is A.

56. "Empowerment" in TQM teams essentially means

(A) the team is given all necessary resources to solve the assigned problem
(B) the team is limited by self-imposed constraints
(C) the team is limited by management imposed constraints
(D) the team must report only to the immediate supervisor of the problem area

Solution:

"Empowerment ... One of the ways this can be done is by structuring work that allows employees to make decisions concerning the improvement of work processes within well-defined parameters." These parameters represent constraints of solutions within employees' spheres of knowledge. (Goetoch, D.L., and Davis, S.B., *Introduction to Total Quality*, Prentice-Hall, 2nd edition, 1997 (pp. 17).)

Answer is C.

57. W. Edward Deming is most noted for

(A) his development of the 14 points an organization should follow to achieve total quality
(B) his contribution to the development of the Malcomb-Baldrige Award
(C) his controversial "point" of eliminating quotas
(D) his great success in implementing TQM in the U.S. steel industry

Solution:

"Over the years, Dr. Deming has developed 14 points that describe what is necessary for a business to survive and be competitive today." (Goetoch, D.L., and Davis, S.B., *Introduction to Total Quality*, Prentice-Hall, 2nd edition, 1997 (pp. 20-21).)

Answer is A.

WORK PERFORMANCE AND METHODS

58. Consider the following work sampling study for a piece of production equipment based upon a random sample of 100 observations.

activity	no. of times observed
machine idle—no product	5
machine operating	85
machine idle—downtime and repair	10

Find the 95% confidence interval for the proportion of the time the machine is idle.

(A) 0.05, 0.25
(B) 0.08, 0.22
(C) 0.10, 0.20
(D) 0.14, 0.16

Solution:

The general formula for finding the 95% confidence interval for a proportion that an activity occurs is

$$p \pm 1.96\sqrt{\frac{p(1-p)}{n}}$$

n = total number of observations

x = times the activity is observed

$p = \frac{x}{n}$

1.96 = standardized normal variant for a 95% confidence interval

For this problem,

$$x = 5 + 10 = 15$$

$$n = 100$$

$$p = \frac{15}{100} = 0.15$$

The confidence interval is

$$0.15 \pm 1.96\sqrt{\frac{(0.15)(0.85)}{100}} = 0.15 \pm 0.07 = 0.08, 0.22$$

Answer is B.

59. The allowed time for a time study is 2.2614 min per unit. Assume workers are allowed two 15 min breaks for personal allowances. Combined allowances for fatigue and unavoidable delay are 15%. How many parts should a worker produce during an 8 hr shift?

(A) 50
(B) 100
(C) 173
(D) 200

Solution:

The standard time is

$$\begin{aligned}&(\text{allowed time})(1 + \text{allowances as a decimal})\\ &= \left(2.2614\ \frac{\text{min}}{\text{unit}}\right)(1.15)\\ &= 2.6006\ \text{min/unit}\end{aligned}$$

The production time is

$$480\ \frac{\text{min}}{\text{day}} - 30\ \frac{\text{min}}{\text{day}} = 450\ \text{min/day}$$

The parts per day a worker should produce is

$$\left(450\ \frac{\text{min}}{\text{day}}\right)\left(\frac{1\ \text{part}}{2.6006\ \text{min}}\right) = 173$$

Answer is C.

60. A coil-slitting operation consists of three elements. A summary of a time study and a work sampling study is shown with all times in minutes.

element	cycles timed	average time	occurrences per cycle	rating
1	35	0.64	1	100
2	35	0.55	1	110
3	35	1.28	1	95

Work Sampling Study

activity	no. of times observed
machine idle—no material available	20
machine idle—maintenance	10
machine working	150
operator away from machine—avoidable	5
operator away from machine—personal	15
	200

Assume there are no official breaks for the employee. What would be the standard time in minutes per unit based upon the results of this study?

(A) 2.4610
(B) 2.7812
(C) 2.8512
(D) 3.0147

Solution:

OT_i = observed average time for element i

n = number of elements

R_i = rating factor

$$\begin{aligned}\text{AT} &= \text{allowed time} = \sum_{i=1}^{n}(\text{OT}_i)\left(\frac{R_1}{100}\right)\\ &= (0.64)(1.00) + (0.55)(1.10) + (1.28)(0.95)\\ &= 2.461\ \text{min}\end{aligned}$$

ST = standard time = (AT)$(1 + A)$

A = allowances

Estimate allowances from the work sampling study.

$$\text{allowance for idle machine} = \frac{20 + 10}{200} = 0.15$$

$$\text{personal allowance} = \frac{15}{200} = 0.075$$

$$A = 0.15 + 0.075 = 0.225$$

$$\text{ST} = (2.461)(1.225) = 3.0147\ \text{min}$$

Answer is D.

Practice Exam

1. A spreadsheet has been developed to calculate the cycle time for an assembly line required to meet a given production demand. Cells A1 through A5 contain the task times, cell B1 contains the number of stations, and cell B2 contains the cycle time. If cell C1 is to contain the balance delay for the solution, what formula should be put into it?

(A) (SUM(A1...A5) + B1*B2)/(B1 + B2)
(B) (B1 − (SUM(A1...A5)/B1*B2))/(SUM(B1...B2))
(C) (B1*B2 − SUM(A1...A5))/SUM(A1...A5)
(D) (B1*B2 − SUM(A1...A5))/B1*B2

2. Ignoring headers, blocks, and line errors, calculate the approximate time to transmit a data file containing 70,000 characters over a 9600 bit/sec transmission line.

(A) 0.01 min
(B) 0.1 min
(C) 1 min
(D) 10 min

3. A data file contains the last 12 months of sales data from the Ohio Valley region. The manager is trying to determine the number of sales personnel to assign to the region for the next month. She has written a pseudocode segment to forecast the expected sales for next month and to assign the number of personnel. The data are

$$400, 500, 600, 400, 800, 850, 950,\\ 1100, 1150, 1300, 1320, 1440$$

The pseudocode segment is as follows.

```
N = 1
Read data point
While N < 8
Increment N by 1
Read data point
Endwhile
Initialize F to 0 and set N = 1
While N < 6
Increment N by 1
Read data point and add value to F
Endwhile
Set P = F/(N − 1)
If P < 1000 set SF = 5
If P > 999 and P < 1300 set SF = 8
If P > 1299 and P < 1450 set SF = 10
Else set SF = 14
```

The number of sales personnel, SF, at the end of this segment would be

(A) 5
(B) 8
(C) 10
(D) 14

4. A two-way factorial design is used to determine whether a treatment is significant at a 0.05 level. The analysis of a variance table from experimental results is shown as follows.

source of variation	degrees of freedom	sum of squares
replications	4	122.46
treatments	4	126.92
residuals	16	116.08

Which of the following statements is true?

(A) One can conclude that the treatment is significant at a 0.05 level.
(B) One can conclude that the treatment is not significant at a 0.05 level.
(C) The replication error is significant, thus one cannot make a conclusion.
(D) The replication error is significant, thus one can conclude that the treatment is significant at a 0.05 level.

5. A 2^2 factorial design with three replications is used to determine significant factors at an 0.05 level. The analysis of the variance table from experimental results is shown as follows.

source of variation	degrees of freedom	sum of squares
factor A	1	108.50
factor B	1	61.22
cross term AB	1	9.60
errors	8	32.00

One can conclude that

(A) factor A is significant at an 0.05 level
(B) factor A and B are both significant at an 0.05 level
(C) factor A and B and cross term AB are all significant at an 0.05 level
(D) none of the factors are significant

6. A 3^k factorial design is used to determine significant factors. However, there are no replications for any combination during experiments. Without conducting any more experiments, the best one can do is

(A) nothing, since there are no error terms
(B) interpolate error terms and reduce one degree of freedom, then conduct the analysis of variance as usual
(C) pool the high-order cross terms as error terms, then conduct the analysis of variance
(D) use the most significant factor as error terms, then conduct the analysis of variance

7. Jane is planning for her retirement. Each month she places \$200 in an account that pays 12% nominal interest, compounded monthly. She makes the first deposit of \$200 on January 31, 1997. The last \$200 deposit will be made on December 31, 2016. If the interest rate remains constant and all deposits are made as planned, the amount in Jane's retirement account on January 1, 2017 is most nearly

(A) \$60,000
(B) \$155,000
(C) \$173,000
(D) \$198,000

8. Dawn purchased a \$10,000 car. She put \$2000 down and financed the \$8000 balance. The interest rate is 9% nominal, compounded monthly, and the loan is to be repaid in equal monthly installments over the next four years. Dawn's monthly car payment is most nearly

(A) \$167
(B) \$172
(C) \$188
(D) \$200

9. James is a major prizewinner in a sweepstakes. He has the option of either receiving a single check for \$125,000 now or receiving a check for \$50,000 each year for three years. (James would be given the first \$50,000 check now.) At what interest rate would James have to invest his winnings for him to be indifferent as to how he receives his winnings?

(A) 15.5%
(B) 20.0%
(C) 21.6%
(D) 23.3%

10. The service time (in hours) for a copy machine is approximately exponentially distributed. In examining the records for 50 breakdowns, it is determined that the average service time is 1.25 hr. An estimate for the probability that a service time will exceed 2 hr is approximately

(A) 0.05
(B) 0.10
(C) 0.15
(D) 0.20

11. You would like to test the null hypothesis at a 5% level of significance that the mean shear strength of spot welds is at least 450 psi. You randomly select 15 welds, measure the shear strength, and determine the following results.

sample mean ($\overline{x}$): 445 psi
sample standard deviation (s): 10 psi

Based upon the data,

(A) the null hypothesis is true
(B) the null hypothesis is false
(C) there is not enough information to say the hypothesis is true
(D) there is not enough information to say the hypothesis is false

12. The p-value is the smallest level of significance that would lead to rejection of the null hypothesis. The p-value in Prob. 11 is approximately

(A) 0.001
(B) 0.005
(C) 0.025
(D) 0.040

13. A part requires that two operations (Op1 followed by Op2) be processed on a single machine. Assume the company will be operating 5 days per week and 10 hours per day with no set-up time between operations. Determine the integer number of machines required to produce 5000 parts per week. The following data is available for the operations.

operation	Op1	Op2
standard time (in minutes)	2.0	4.0
actual performance	0.90	0.90
machine availability	0.90	0.95
scrap rate	0.03	0.05

(A) 10
(B) 12
(C) 13
(D) 15

14. Six assembly cells (each 35 m × 35 m) are to be located in an area that will configure into a 2 cell × 3 cell facility. The cost of material handling is directly proportional to the distance traveled from centroid to centroid. The following material flow data is given.

cell	C1	C2	C3	C4	C5	C6
C1	–	150		300	0	50
C2		–	50	0	10	80
C3			–	150	100	30
C4				–	40	0
C5					–	100
C6						–

If the current cell layout is shown in the figure below, what would be the impact on the total material handling flow if C2 and C4 were to exchange positions?

C1	C3	C4
C6	C5	C2

(A) −1200 (decrease)
(B) 0 (no change)
(C) 2000 (increase)
(D) 3600 (increase)

15. Consider a warehouse that contains 40 storage locations. Four products are stored in the warehouse. Storage and throughput requirements are given as follows for each product. In what order (or priority) should products be assigned to storage locations to minimize the total distance traveled?

product	no. of locations required (S)	total loads moved per day (T)
1	10	200
2	16	160
3	8	240
4	6	150

(A) 3, 4, 1, 2
(B) 3, 1, 2, 4
(C) 2, 1, 3, 4
(D) 2, 1, 4, 3

16. In the design of an automobile radiator, an engineer has a choice of using either a brass-copper alloy casting or a plastic molding. Either material provides the same service. However, the brass-copper alloy casting has a mass of 25 lbm, compared with 16 lbm for the plastic molding. Every pound of extra mass in the automobile has been assigned a penalty of \$4 to account for increased fuel consumption during the lifecycle of the car. The brass-copper alloy casting costs \$3.35 per pound, whereas the plastic molding costs \$7.40 per pound. Machining costs per casting are \$6.00 for the brass-copper alloy. Which material should the engineer select, and what is the difference in unit cost?

(A) select brass-copper alloy; savings = \$1.35/radiator
(B) select brass-copper alloy; savings = \$6.00/radiator
(C) select plastic; savings = \$7.35/radiator
(D) select plastic; savings = \$8.50/radiator

17. A process engineer is trying to decide on the type of tool material to use for a machining operation. Relevant data for the alternatives are as follows.

	tool material A	tool material B
tool cost:	\$100	\$30
production rate:	100 pieces/hr	75 pieces/hr
tool life:	50 hr	25 hr

It takes 1 hour to replace a worn tool. Labor costs \$18 per hour, and a total of 15,000 pieces are needed. What tool material should be selected, and what are the expected savings?

(A) material A; save \$930
(B) material B; save \$1020
(C) material A; save \$1020
(D) material B; save \$3054

18. A project engineer is deciding on the most cost-effective duration for a new project. The direct costs of the project are expected to vary indirectly with project duration, while indirect costs vary directly with the square of the project duration. The total project cost is represented by the following equation.

$$\text{project cost} = \$5000 + \frac{\$4000 \text{ months}}{X} + \frac{\$250X^2}{\text{months}^2}$$

X = project duration in months

For what duration should the project be planned, and what is the expected project cost?

(A) 3 months; $583
(B) 2 months; $3000
(C) 2 months; $8000
(D) 3 months; $8583

19. A cart designed for pushing is 159.6 cm in height. Assuming that height is normally distributed, what percent of women pushing the cart will be able to see over the top? Assume that shoe height is 3 cm.

(A) about 5%
(B) about 8%
(C) about 11%
(D) about 21%

20. A chair with armrests is being designed. Assume a clothing allowance of 1.2 cm. What should be the distance between armrests to accommodate 95% of the population?

(A) 25 cm
(B) 33 cm
(C) 38 cm
(D) 45 cm

21. Consider the following lifting operation that occurs throughout an 8 hr shift. The operation is to lift a casting that is located on a pallet an average of 12 in off the floor and 16 in in front of the operator. The weight is lifted vertically an average total distance of 48 in. This lift occurs an average of 1 lift every 2.5 min. The maximum frequency of lifting that can be sustained for this job is 12 lifts/min. Using the NIOSH formula, what is the maximum permissible limit for the weight of the casting?

(A) 40
(B) 55
(C) 62
(D) 80

22. Measuring industrial performance remains a difficult task. To address this issue, the American Productivity Center has introduced a powerful integrated measurement criterion for measuring total productivity. It has the following format.

$$\text{total productivity} = (\text{productivity})(\text{price recovery})$$

$$\text{productivity} = \frac{\text{percent change in output quantity}}{\text{percent change in resource quantity}}$$

$$\text{price recovery} = \frac{\text{percent change in output price}}{\text{percent change in resource cost}}$$

The data in the following table was taken from an automotive parts supplier.

		period (n)	period ($n+1$)
output	units produced	1200	1400
	price per unit ($)	54.08	57.23
resource	hours worked	8200	7800
	cost per hour ($)	12.05	13.15

The firm's overall productivity improvement from period n to period $n+1$ is most nearly

(A) 1.8%
(B) 12%
(C) 19%
(D) 122%

23. In the late 1970s and early 1980s, many U.S. companies wished to copy the successful management practices of their Japanese competitors. Initially, one of the most widely adopted Japanese management practices was that of quality circles (QCs). However, studies have shown a very high failure rate of QCs in U.S. companies.

Recent research has shown that the primary cause of the high failure rate is

(A) that the U.S. companies adopted the QC concept as a single technique while, in fact, in Japan QCs are just one part of a larger system
(B) the extreme cultural differences between Japanese and U.S. workers
(C) the misunderstanding and improper application of the details of QC practices by the U.S. companies
(D) that U.S. companies are much more highly unionized than their Japanese counterparts

24. Many Japanese management concepts have migrated to the U.S. in recent years and many Japanese phrases and names of Japanese industrial pioneers are now commonly used in U.S. business.

Which of the following phrases/names is not correctly defined?

(A) Taguchi—a simplified system for experimental design, resulting in what some call robust design
(B) Ishikawa—a quality control stratification technique, resulting in what some call fish-bone diagrams

(C) Shingo—a recognition that all companies compete on one of four strategic dimensions (cost, quality, flexibility, or delivery), resulting in what some call the value-chain system
(D) Kanban—a system of simple work flow control techniques, which some call card-based inventory control

25. If the alphabetic characters a, b, and c have ASCII (American Standard Code for Information Interchange) values of 97, 98, and 99, respectively, which character(s) has (have) been transmitted if we receive a digital signal corresponding to the binary string 1100001_2?

(A) character a
(B) character b
(C) character c
(D) characters ac

26. In computer serial communication using modems over a telephone line,

(A) analog data is transmitted directly as an analog signal with amplification
(B) analog data is multiplexed into an analog signal on multiple channels
(C) digital data is transmitted as a digital signal using a differentiated coding method
(D) digital data is modulated onto an analog carrier wave using shift-keying methods

27. Computer-based data communications within organizations have proliferated primarily as the result of

(A) point-to-point direct communications technology
(B) Open Systems Interconnection (OSI) seven-layer communications architecture
(C) multimedia world wide web browser client-server communications
(D) compact disc (CD) technology and its data compression features

28. A single-point tool is used on an engine lathe to make a single pass on a 2.000 in diameter (diameter before the cut), 24 in long part. The depth of cut is set to 0.125 in at a spindle speed of 400 rev/min with a feed of 0.012 in/rev.

The machining time required for a single pass is most nearly

(A) 0.21 min
(B) 5.0 min
(C) 10 min
(D) 115 min

29. A short-stroke reciprocating saw can cut ductile steel at about 12 in^2/min. The operator can unclamp, clamp, and advance stock in the saw on an average of 15 sec per cut. The saw is currently loaded with 3 in diameter low-carbon steel bar stock.

Approximately how long will it take to cut twelve 4 in slugs?

(A) 0.21 min
(B) 5.0 min
(C) 10 min
(D) 115 min

30. The material and manufacture of standard beverage cans have changed dramatically over the past 25 years. Originally steel, nearly all cans are now made of aluminum. The processes used to form the aluminum cans has also been refined and improved.

Which of the following processes is not widely used in the production of today's standard beverage cans?

(A) deep drawing
(B) blanking
(C) crimping
(D) seam welding

31. The wheel assembly for a rolling cart requires 11 components (wheels, nuts, bolts, washers, and the yoke), two hand tools (a screwdriver and pliers), and two assembly directions (from the side and the bottom). Which of the following designs for manufacturing/design for assembly guidelines are appropriate for improving this product?

rule 1: Avoid flexible components that are difficult to handle.

rule 2: Use the simplest possible operations.

rule 3: Reduce the number of components.

rule 4: Exaggerate asymmetry when not possible to provide symmetry.

rule 5: Use snap fasteners if possible.

(A) rules 1 and 5
(B) rules 1, 2, and 4
(C) rules 3 and 5
(D) rules 2, 3, and 4

32. The sculptured surface shown by a wire mesh can be modeled and manufactured on a 5-axis milling machine more precisely as a

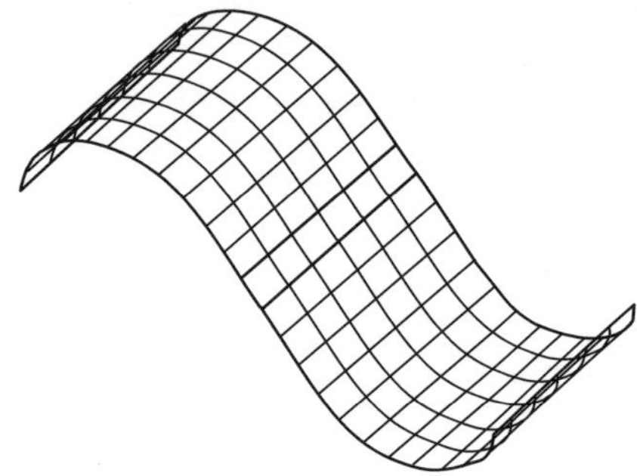

(A) boundary representation prismatic model
(B) $2^1/_2$-dimension vertical milling contour model
(C) polyhedral constructive solid geometry model
(D) nonuniform rational B-spline surface model

33. If a blind circular pocket with a flat bottom is to be machined on a numerically controlled vertical milling machine, the most appropriate machining specification in a numerical control code-generating program will be

(A) end mill in a zigzag tool path with a right offset from the pocket boundary in a counterclockwise direction
(B) ball-nose end mill in a zigzag path with a left offset from the pocket boundary in a clockwise direction
(C) end mill in a spiral tool path with a left offset from the pocket boundary in a counterclockwise direction
(D) ball-nose end mill in a spiral tool path with a right offset from the pocket boundary in a counterclockwise direction

34. Consider an open-loop conveyor system design problem where parts arrive at a 1.2 m wide conveyor input point at a rate of 100 parts per minute. Five parts are first put on a 1 m × 1.3 m pallet, then the pallets are transported on the conveyor. The distance from the loading to unloading point is 200 m. Calculate the minimum speed of the conveyor to support the specified material flow.

(A) 20 m/min
(B) 22 m/min
(C) 26 m/min
(D) 30 m/min

35. A production system requires 40 material-handling moves per hour, with each move having an average distance of 240 m. Each move requires a 1 min load and a 1 min unload. Four forklifts are used to move the loads, and each travels at an average of 80 m/min. The average utilization for the four forklifts is most nearly

(A) 50%
(B) 70%
(C) 85%
(D) 95%

36. A company produces two products that use a combination of six production processes arranged into three cells. Due to the size of the products, a forklift has been selected as the material handling equipment. What is the intercellular material handling flow associated with the following cellular layout? Assume material handling cost is proportional to distance.

	routing						loads
product 1	P1	P2	P3	P4	P5	P6	100
product 2	P2	P5	P4	P3	P1	P6	200

distance between cells in meters (processes in cell)	cell 1 (P1, P2)	cell 2 (P3, P4)	cell 3 (P5, P6)
C1 (P1, P2)	–	50	100
C2 (P3, P4)		–	50
C3 (P5, P6)			–

(A) 6500
(B) 27,500
(C) 51,000
(D) 70,000

37. For $f(x, y) = y^2 + 4xy + 2x^2 - 4x - 2y$, what point gives the minimum value?

(A) There is no minimum value. The function is concave and has only a maximum.
(B) $x = 0$ and $y = 0$
(C) There is no minimum value. There is only an inflection point.
(D) $x = 0$ and $y = 1$

38. Consider the problem of assigning four operators to four machines. The assignment costs in dollars are given below. Operator 1 cannot be assigned to machine 2, and operator 4 cannot be assigned to machine 4. What is the minimum cost assignment for this problem?

	machine 1	2	3	4
operator 1	20	–	20	8
operator 2	14	16	8	12
operator 3	36	12	20	28
operator 4	28	8	24	–

(A) $36
(B) $48
(C) $50
(D) $68

39. Consider a transportation problem that has the following supplies and demands, and cost matrix C where the rows represent the sources and the columns represent the destinations.

source	1	2	3	4
supply	20	40	10	60

destination	1	2	3	4	5
demand	5	15	40	35	35

C	destination				
source	10	15	5	3	12
	11	4	12	23	19
	9	13	24	28	21
	23	34	42	51	29

The objective function value of the solution generated by the northwest corner rule would be

(A) 85
(B) 130
(C) 2575
(D) 3325

40. Jobs go through a painting station and then an automatic inspection station. Each station can only service one job at a time. Four jobs are currently awaiting processing. The time each job must spend at each station is shown as follows.

job	painting time (min)	inspection time (min)
1	12	9
2	8	5
3	15	8
4	7	4

The makespan of the job sequence $\{3, 2, 1, 4\}$ is

(A) 41 min
(B) 44 min
(C) 46 min
(D) 48 min

41. Component Y is used in several products assembled by company ABC. The average demand for component Y is 750 units per week. The fixed cost of placing an order for component Y is $15. The inventory holding cost for component Y is $0.10 per unit per week. To minimize the total cost of ordering and inventory holding, component Y would be most economically ordered in quantities of

(A) 100 units
(B) 316 units
(C) 475 units
(D) 750 units

42. The demand for product Z is 10,000 units per week. Each unit of product Z requires 0.25 labor hours to produce. There are two 8.5-hour production shifts per day, five days per week. During each shift, approximately 1.5 hours are nonproductive (due to breaks, setup, etc.). The minimum workforce needed to meet the demand for product Z without requiring overtime is

(A) 35 people/day
(B) 36 people/day
(C) 35 people/shift
(D) 36 people/shift

43. Which of the following contains only those tools typically used for conducting a graphical analysis of worker productivity?

(A) operation process charts, cash flow diagrams, and critical path diagrams
(B) value-added charts, flow diagrams, and force/stress diagrams
(C) cash flow diagrams, PERT diagrams, and flow process charts
(D) operation process charts, flow process charts, and left-hand/right-hand charts

44. An assembly department has been complaining that it cannot meet its production standard due to the amount of time it has to spend finding missing parts. To determine the percentage of time that an assembly department spends looking for missing parts, an industrial engineer decides to conduct a work sampling study. If the department has been reporting that 20% of its time is used looking for missing parts, how many observations must the industrial engineer make in order to prove the department's claim with a 90% confidence level, ±5%?

(A) 43
(B) 121
(C) 1732
(D) 4303

45. A plant manager has been asked by the company board of directors to provide a summary of the plant's productivity gains over the past 5 years. Though there are many ways in which productivity can be measured, this company uses the variable cost per unit produced. Data on the company's performance over the past 5 years is given in the following table.

	year 1	year 2	year 3	year 4	year 5
total variable cost	\$150,000	\$170,000	\$180,000	\$210,000	\$220,000
total units produced	120,000	150,000	155,000	195,000	210,000

The average change in variable cost per unit over the past 5 years is most nearly

(A) 4%
(B) 8%
(C) 10%
(D) 16%

46. Consider two stocks. Stock 1 always sells for either \$20 or for \$40. If stock 1 is selling for \$20 today, there is a 70% chance that it will sell for \$20 tomorrow. If it is selling for \$40 today, there is an 85% chance it will sell for \$40 tomorrow. Stock 2 always sells for either \$20 or \$35. If stock 2 is selling for \$20 today, there is a 90% chance that it will sell for \$20 tomorrow. If it is selling for \$35 today there is an 80% chance it will sell for \$35 tomorrow. What is the expected value of stock 2?

(A) \$20
(B) \$25
(C) \$28
(D) \$46

47. The final machining of a product requires three different stages. The stages of the machining are consecutive. Each stage consists of a single machine doing a specific task. The machining time at each stage is exponentially distributed with a mean of 20 min. A product must wait to begin the machining process until the current product has completed stage 3. Products arrive at stage 1 according to a Poisson process at the rate of two per hour, and the facility has a buffer that can hold at most ten of these products as they wait for machining. Using Kendall's notation, what type of queuing system is this?

(A) M/M/3/10
(B) M/M/10/3
(C) M/G/1/3/10
(D) M/E_3/1/11

48. A university's food court has two tellers. People waiting to pay for their food form a single line and go to the first available teller. There is room in the food court for 200 people, and there is also room outside to wait. Customers have an interarrival time to the food court that is exponentially distributed with a mean of 2 min. The two tellers work at the same rate. The service time for each teller is exponentially distributed with a mean of 3 min. On average, the number of people waiting in line for the tellers is most nearly

(A) 0
(B) 1
(C) 2
(D) 30

49. When running a simulation experiment, data are gathered on output Y. For each replication of the experiment, a single statistic, $\overline{Y}$ (the average Y), is saved. When analyzing the data across all replications, statistical tests are applied to an aggregated statistic, $\overline{\overline{Y}}$ (the average of the $\overline{Y}$).

When it is desired to perform statistical tests, the reason for analyzing experimental statistics in this way is to take advantage of what statistical phenomenon?

(A) Chebyshev's theorem
(B) central limit theorem
(C) Markov's theorem
(D) Kolmogorov's theorem

50. Simulation experiments are often repeated in what are called runs or, more commonly, replications.

Replications are used in simulation experimentation because

(A) random numbers generated by computer simulation packages are not truly random, and so the sequences of numbers they produce repeat after only a small quantity of numbers are drawn
(B) many standard statistical tests used to analyze simulation outputs assume the data to be independent and identically distributed, and statistics gathered within a single replication rarely meet those requirements
(C) bias due to the initial startup conditions must be recognized, and replication is an effective method to eliminate such bias
(D) most simulation packages have difficulty storing the vast amounts of data that would need to be collected during a single long run if no replications were used

51. A simulation model was run for 30 replications and the mean utilization of a transporter was recorded for each replication. Those 30 data points were then used to form a confidence interval on mean transporter utilization for the system. At 95% confidence level, that interval was found to be $37.2\% \pm 3.4\%$.

Given this information, which of the following facts can be definitively stated about the system?

(A) At 95% confidence, the 30-replication sample mean of transporter utilization lies in the range $37.2\% \pm 3.4\%$.
(B) At 95% confidence, the population mean of transporter utilization lies in the range $37.2\% \pm 3.4\%$.
(C) At 95% confidence, the transporter never reached 100% utilization at any point during any of the 30 replications.
(D) With only 30 replications, no definitive statements can be made about the system, including the utilization of the transporter.

52. If a lot is submitted from a process that has a population mean proportion defective of 30%, the probability of accepting this lot is most nearly

(A) 0%
(B) 10%
(C) 40%
(D) 75%

53. 25 samples of size 5 were drawn from a population, and the average range for a measured diameter was determined to be 8.960. What is the estimate of the population standard deviation?

(A) 3.850
(B) 4.961
(C) 6.000
(D) 8.956

54. The permanent control chart limits for an R-chart used to track the sample ranges described in Prob. 53 should be set at most nearly

(A) 18.910; 0
(B) 18.941; 0
(C) 19.810; 0
(D) 19.905; 0

55. Considering that costs of training company personnel at all levels could be significant, who would you advise (in a company of 500 employees) to take a training course on the tools and concepts of TQM?

(A) CEO and upper management
(B) CEO, upper management, and middle management
(C) CEO, upper management, middle management, and front line supervisors
(D) CEO, upper management, middle management, front line supervisors, and line operators

56. The Malcolm-Baldrige Award Program

(A) is a relatively easy program to implement into a manufacturing/service company
(B) is available only to manufacturing companies
(C) evaluates competitors according to a seven-category criteria
(D) is a worldwide competition

57. Quality circles is a technique

(A) identical to a "brainstorming program"
(B) wherein the leader is typically a manager
(C) wherein members are selected by management to solve a specific problem
(D) wherein members volunteer to identify and recommend workplace improvements

58. Consider the following work sampling study for a piece of production equipment based upon a random sample of 100 observations.

	number of times observed
machine idle—no product:	5
machine operating:	85
machine idle—downtime and repair:	10

Suppose that an accuracy of 0.05 is required for the study with a confidence of 95%. How many additional observations must be made?

(A) 0
(B) 50
(C) 96
(D) 150

59. Consider the following time study summary for a particular job. Element 2 is a machine-controlled element. Element 4 occurs every third cycle. All times are in minutes.

element	cycles timed	average	standard deviation	occurrences per cycle	rating
1	30	0.246	0.031	1	110
2	30	1.214	0.001	1	100
3	30	0.252	0.022	1	110
4	10	1.682	0.015	$\frac{1}{3}$	90

What is the allowed time, in minutes, for this job before allowances?

(A) 1.8423
(B) 2.2664
(C) 3.1122
(D) 3.2120

60. Consider the time study summary for Prob. 59.

An accuracy of ± 0.01 min is required for each element. How many additional cycles must be timed for 95% accuracy?

(A) 0
(B) 7
(C) 10
(D) 50

SOLUTIONS FOR THE PRACTICE EXAM

1. The balance delay is

$$\frac{(\text{no. of stations})(\text{cycle time}) - \text{sum of task times}}{(\text{no. of stations})(\text{cycle time})}$$

Answer is D.

2.

$$t = \text{transmission time}$$
$$f = \text{number of bits to be transmitted}$$
$$r = \text{transmission line rate}$$
$$t = \frac{f}{r}$$

Assuming a character is represented by 8 bits, then

$$f = (70{,}000)(8 \text{ bits})$$
$$t = \frac{(70{,}000)(8 \text{ bits})}{9600 \ \frac{\text{bits}}{\text{sec}}} = 58.3 \text{ sec} \quad (1 \text{ min})$$

Answer is C.

3. The first "while" statement in the pseudocode reads the first seven data points from the data file. The second "while" statement reads the next five data points from the data file and adds them together. It also increments N by 1 each time. This yields the value $F = 1100 + 1150 + 1300 + 1320 + 1440 = 6310$ and $N = 6$. The code then sets $P = F/(N-1)$, which gives $P = 6310/5 = 1262$. The value 1262 satisfies the second "if" statement, hence the value for SF is 8.

Answer is B.

4. From the table,

$$F^*_{4,16,0.05} = 3.01$$

The mean squares for errors is

$$\frac{116.08}{16} = 7.255$$

The mean squares for replications is

$$\frac{122.46}{4} = 30.62$$
$$F = \frac{30.62}{7.255} = 4.22$$

The mean squares for treatments is

$$\frac{126.92}{4} = 31.73$$
$$F = \frac{31.73}{7.255} = 4.37$$

Although the treatments are significant at 0.05, the replications are also significant. The results have a lack-of-fit problem, and a conclusion cannot be made.

Answer is C.

5. From the table,

$$F^*_{1,8,0.05} = 5.32$$

The mean squares for errors is

$$\frac{32}{8} = 4.00$$

The mean square for A is

$$\frac{108.5}{1} = 108.50$$
$$F = \frac{108.50}{4.00} = 27.13$$

The mean square for B is

$$\frac{61.22}{1} = 61.22$$
$$F = \frac{61.22}{4.00} = 15.31$$

The mean square for AB is

$$\frac{9.60}{1} = 9.60$$
$$F = \frac{9.60}{4.00} = 2.40$$

F values from factors A and B are greater than F^*. Thus, factors A and B are both significant at a 0.05 level.

Answer is B.

6. Choice (C) is correct. The rationale is that for most engineering properties, the high-order cross terms are usually nonexistent, thus they can be used as error terms if more experiments are not feasible.

Answer is C.

7. Use the uniform series compound amount factor to find the future worth of the deposits.

$$F = A(F/A, i\%, n)$$

A = uniform series of end of compounding period cash flows = \$200

$$i = \text{interest rate per compounding period} = \frac{12\%}{12 \text{ compounding periods per year}} = 1\% \text{ per compounding period (month)}$$

$$n = \text{number of compounding periods} = (20 \text{ years})(12 \text{ compounding periods per year}) = 240 \text{ compounding periods}$$

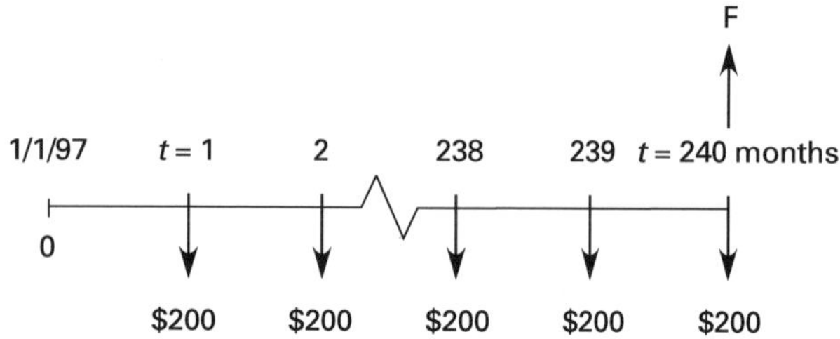

F = future equivalent value of a cash flow or series of cash flows

$$= (\$200)(F/A, 1\%, 240) = (\$200)\left[\frac{(1+0.01)^{240}-1}{0.01}\right] = \$197{,}851 \quad (\$198{,}000)$$

Answer is D.

8. Use the capital recovery factor.

$$A = P(A/P, i\%, n)$$

P = present equivalent value of a cash flow or series of cash flows (loan amount) = \$8000

$$i = \text{interest rate per compounding period} = \frac{9\%}{12 \text{ compounding periods per year}} = 0.75\% \text{ per compounding period}$$

$$n = \text{number of compounding periods} = (4 \text{ years})(12 \text{ compounding periods per year}) = 48 \text{ compounding periods}$$

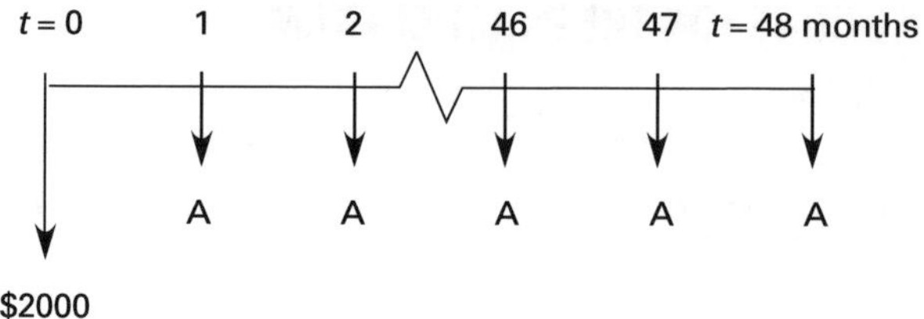

A = uniform series of end of period cash flows

$$A = (\$8000)(A/P, 0.75\%, 48) = (\$8000)\left[\frac{(0.0075)(1+0.0075)^{48}}{(1+0.0075)^{48}-1}\right] = \$199.20 \quad (\$200)$$

Answer is D.

9. Equate the equivalent worth of the two options and solve for the interest rate.

Option 1: Receive \$125,000 now (P_0 = \$125,000).

Option 2: Receive \$50,000 now and \$50,000 each year for two years.

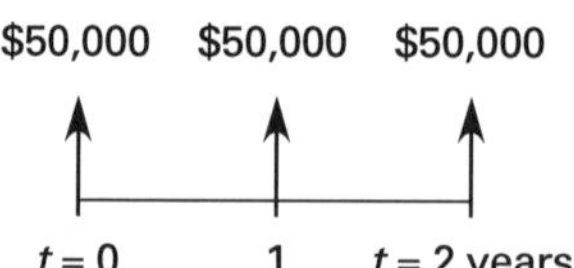

Use the uniform series present worth factor.

$$P_0 = \$50{,}000 + A(P/A, i\%, n)$$
$$A = \$50{,}000$$
$$n = 2$$
$$P_0 = \$50{,}000 + (\$50{,}000)(P/A, i\%, 2) = \$50{,}000 + (\$50{,}000)\left[\frac{(1+i)^2-1}{i(1+i)^2}\right]$$

Equate the present worth of the options and solve for the interest rate.

$$\$125{,}000 = \$50{,}000 + (\$50{,}000)\left[\frac{(1+i)^2-1}{i(1+i)^2}\right]$$
$$1.5 = \left[\frac{(1+i)^2-1}{i(1+i)^2}\right] = (P/A, i\%, 2)$$

By trial and error, a range for i can be determined. Then use linear interpolation to approximate i.

interest rate	$(P/A, i\%, 2)$
20	1.5278
i	1.5000
25	1.4400

$$\frac{25\% - i\%}{1.44 - 1.5} = \frac{25\% - 20\%}{1.44 - 1.5278}$$
$$i = 0.216 \quad (21.6\%)$$

Answer is C.

10. Let t = service time in hours. $f(t)$ is exponentially distributed.

$$f(t) = \lambda e^{-\lambda t} \quad [t > 0]$$
$$\lambda = \frac{1}{\mu}$$
$$\mu = \text{the population mean}$$

Let the sample mean be an estimate of the population mean.

The estimate for μ is the sample average, which is 1.25.

$$\lambda = \frac{1}{1.25} = 0.8$$
$$P(t > 2) = \int_2^{\infty} 0.8e^{-0.8t}dt = -e^{-0.8t}\Big|_2^{\infty}$$
$$= 0 + e^{-1.6} = 0.2018 \quad (0.20)$$

Answer is D.

11.
$$H_0\colon \mu \geq 450$$
$$H_1\colon \mu < 450$$

Accept hypothesis H_0 if

$$\frac{\overline{x} - 450}{\frac{s}{\sqrt{15}}} \geq -t_{0.05,14} = -1.761$$

$$t = \frac{445 - 450}{\frac{10}{\sqrt{15}}} = -1.936$$
$$-1.936 < -1.761$$

Reject H_0.

Answer is B.

12.
$$t = \frac{\overline{x} - 450}{\frac{s}{\sqrt{15}}} = -1.936$$

$$p[t_{n=14} < -1.936] = x$$

By interpolation,

$$0.05 \longrightarrow -1.761$$
$$x \longrightarrow -1.936$$
$$0.025 \longrightarrow -2.145$$

$$x = 0.05 - \left(\frac{0.175}{0.384}\right)(0.025) = 0.0386 \quad (0.040)$$

Answer is D.

13. F = total number of machines ceiling $= \sum F_k$

$$= F_1 + F_2$$
$$F_k = \frac{S_{i,k}Q_i}{EHR\prod_i(1 - P_{i,k})}$$
$$= \text{number of machines for operation } k$$

S_i = standard time required per part i for operation k

Q_i = production quantity per time period for part i

E = actual performance of machine

H = available time units per time period

R = machine k availability

$P_{i,k}$ = scrap rate for machine k part i

$$F_1 = \frac{(2)(5000)}{(0.9)\big((5)(10)(60)\big)(0.9)(1 - 0.03)(1 - 0.05)}$$
$$= 4.47$$
$$F_2 = \frac{(4)(5000)}{(0.9)\big((5)(10)(60)\big)(0.95)(1 - 0.05)}$$
$$= 8.21$$

$$F = \text{ceiling } (4.47 + 8.21) = \text{ceiling } (12.68) = 13$$

Answer is C.

14. To calculate the total material handling cost impact, $\Delta\text{TC}_{2,4}$, it is necessary to calculate the change in cost resulting from exchanging the two cells (C2 and C4). Let w be the weight or amount of flow between cells and d be the distance between cells.

$$\begin{aligned}\Delta TC_{2,4} &= (w_{1,2} - w_{1,4})(d_{1,2} - d_{1,4}) \\ &\quad + (w_{2,2} - w_{2,4})(d_{2,2} - d_{2,4}) \\ &\quad + (w_{3,2} - w_{3,4})(d_{3,2} - d_{3,4}) \\ &\quad + (w_{4,2} - w_{4,4})(d_{4,2} - d_{4,4}) \\ &\quad + (w_{5,2} - w_{5,4})(d_{5,2} - d_{5,4}) \\ &\quad + (w_{6,2} - w_{6,4})(d_{6,2} - d_{6,4}) \\ &\quad - 2w_{2,4}(d_{2,4}) \\ &= (150 - 50)(30 - 10) + (0 - 80)(0 - 20) \\ &\quad + (50 - 30)(20 - 20) + (0 - 0)(10 - 30) \\ &\quad + (10 - 100)(10 - 10) + (80 - 0)(20 - 0) \\ &\quad - (2)(80)(20) \\ &= 2000 + 1600 + 0 + 0 + 0 + 1600 - 3200 \\ &= 2000\end{aligned}$$

Answer is C.

15. Products are assigned locations based on their throughput (T) to number of location (S) ratio (T/S). The order (or priority) is by non-increasing T/S ratio.

product	no. of locations required (S)	total loads moved per day (T)	T/S
1	10	200	$\frac{200}{10} = 20$
2	16	160	$\frac{160}{16} = 10$
3	8	240	$\frac{240}{8} = 30$
4	6	150	$\frac{150}{6} = 25$

Non-increasing order of T/S is 3, 4, 1, 2.

Answer is A.

16.

cost factor (piece)	brass-copper alloy	plastic molding
casting	(25 lbm)($3.35/lbm) = $83.75	(16 lbm)($7.40/lbm) = $118.40
machining	$6.00	0.00
weight penalty	(25 lbm − 16 lbm)($4/lbm) = $36.00	0.00
total cost	$125.75	$118.40

The plastic molding should be selected to save $125.75 − $118.40 = $7.35 over the lifecycle of each radiator.

Answer is C.

17. Compute the cost of producing 15,000 pieces for each tool material.

$$\text{total cost} = (\text{tool cost})(\text{no. of tools needed}) + \left(\frac{\text{labor cost}}{\text{hr}}\right)(\text{hr needed})$$

$$\text{tools needed} = \frac{\text{pieces required}}{\left(\frac{\text{pieces}}{\text{hr}}\right)\left(\frac{\text{tool life in hr}}{\text{tool}}\right)}$$

$$\begin{aligned}\text{hours needed} &= \text{production time} + \text{tool changing time} \\ &= \frac{\text{pieces required}}{\frac{\text{pieces}}{\text{hr}}} + \left(1\ \frac{\text{hr}}{\text{tool}}\right)(\text{no. of tools needed})\end{aligned}$$

For material A,

$$\text{tools needed} = \frac{15{,}000 \text{ pieces}}{\left(100\ \frac{\text{pieces}}{\text{hr}}\right)\left(50\ \frac{\text{hr}}{\text{tool}}\right)} = 3 \text{ tools}$$

$$\text{hours needed} = \frac{15{,}000 \text{ pieces}}{100\ \frac{\text{pieces}}{\text{hr}}} + \left(1\ \frac{\text{hr}}{\text{tool}}\right)(3 \text{ tools}) = 153 \text{ hr}$$

$$\text{total cost} = \left(\frac{\$100}{\text{tool}}\right)(3 \text{ tools}) + \left(\frac{\$18}{\text{hr}}\right)(153 \text{ hr}) = \$3054$$

For material B,

$$\text{tools needed} = \frac{15{,}000 \text{ pieces}}{\left(75\ \frac{\text{pieces}}{\text{hr}}\right)\left(25\ \frac{\text{hr}}{\text{tool}}\right)} = 8 \text{ tools}$$

$$\text{hours needed} = \frac{15{,}000 \text{ pieces}}{75\ \frac{\text{pieces}}{\text{hr}}} + \left(1\ \frac{\text{hr}}{\text{tool}}\right)(8 \text{ tools}) = 208 \text{ hr}$$

$$\text{total cost} = \left(\frac{\$30}{\text{tool}}\right)(8 \text{ tools}) + \left(\frac{\$18}{\text{hr}}\right)(208 \text{ hr}) = \$3984$$

Material A should be selected. The savings are $3984 − $3054 = $930.

Answer is A.

18. To minimize project cost, set the first derivative of the project cost equation (with respect to project length) equal to zero and solve for the project length, X.

$$\frac{d(\text{project cost})}{dX} = -\frac{\$4000 \text{ months}}{X^2} + \frac{\$500X}{\text{months}^2} = 0$$

$$\$500X^3 = \$4000 \text{ months}^3$$

$$X = \sqrt[3]{8 \text{ months}^3} = 2 \text{ months}$$

To find the expected project cost, substitute $X = 2$ months into the project cost equation.

$$\begin{aligned}\text{project cost} &= \$5000 + \frac{\$4000 \text{ months}}{X} + \frac{\$250X^2}{\text{months}^2} \\ &= \$5000 + \frac{\$4000 \text{ months}}{2 \text{ months}} \\ &\quad + \frac{(\$250)(2 \text{ months})^2}{\text{months}^2} \\ &= \$8000\end{aligned}$$

Answer is C.

19. x = eye height from floor (cm)

From an ergonomics table, mean eye height is

$$148.9 \text{ cm} + 3 \text{ cm} = 151.9 \text{ cm}$$

The standard deviation is 6.4 cm.

$$\begin{aligned}P[x > 159.6] &= P\left[z > \frac{159.6 - 151.9}{6.4}\right] \\ &= P(z > 1.20) \\ &= 0.1151 \quad (11\%)\end{aligned}$$

Answer is C.

20. Find hip breadth while sitting for the 95th percentile from an ergonomics table.

women: 43.7 cm
men: 40.6 cm

Design for women.

$$43.7 \text{ cm} + 1.2 \text{ cm} = 44.9 \text{ cm} \quad (45 \text{ cm})$$

Answer is D.

21. The action limit is

$$\begin{aligned}\text{AL} &= (90)\left(\frac{6}{H}\right)(1 - 0.01(|V - 30|)) \\ &\quad \times \left(0.7 + \frac{3}{D}\right)\left(1 - \frac{F}{F_{\text{max}}}\right)\end{aligned}$$

H = horizontal distance of the hand from the body's center of gravity at the beginning of the lift
= 16 in

V = vertical distance from the hands to the floor at the beginning of the lift
= 12 in

D = distance the object is lifted vertically
= 48 in

F = average number of lifts per minute
= 0.4/min

F_{max} = maximum frequency of lifting that can be sustained over an 8 hr shift
= 12 lifts/min

MPL = maximum permissible limit
= 3 (action limit)

$|V - 30|$ = absolute value of $V - 30$

$$\begin{aligned}\text{AL} &= \left(\frac{(90)(6)}{16}\right)(1 - 0.01(|12 - 30|)) \\ &\quad \times \left(0.7 + \frac{3}{48}\right)\left(1 - \frac{0.4}{12}\right) \\ &= 20.4\end{aligned}$$

$$\text{MPL} = (20.4)(3) = 61.2 \quad (62)$$

Answer is C.

22. total productivity

$$\begin{aligned}&= (\text{productivity})(\text{price recovery}) \\ &= \left(\frac{\text{percent change in output quantity}}{\text{percent change in resource quantity}}\right) \\ &\quad \times \left(\frac{\text{percent change in output price}}{\text{percent change in resource cost}}\right) \\ &= \left(\frac{\frac{1400 \text{ units}}{1200 \text{ units}}}{\frac{7800 \text{ hr}}{8200 \text{ hr}}}\right)\left(\frac{\frac{\$57.23}{\$54.08}}{\frac{\$13.15}{\$12.05}}\right) \\ &= 1.189\end{aligned}$$

There is an 18.9% (19%) total productivity improvement.

Answer is C.

23. Recent examinations of U.S. companies, both unionized and nonunionized, have shown significant success with Japanese management techniques when these techniques are adopted as a total system. Implementing Japanese techniques piecemeal has been, for the most part, unsuccessful, while implementing overall Japanese management systems has been significantly more successful.

Answer is A.

24. Shigeo Shingo is a Japanese production systems expert widely known for his concept of reducing set-up times by recognizing the differences between what he calls internal setup (which can only be done while the machine is idle) and external setup (which can be performed while the machine is operating). His concepts are called Single Minute Exchange of Dies, or SMED.

Answer is C.

25. The binary string converts to the decimal string 97, which is the ASCII value for alphabetic character a. The calculation is

$$\text{decimal value} = \sum p_i(2^i) \quad [p_i \text{ is the binary value}]$$

$$\begin{aligned}\text{decimal equivalent} &= (1\times 2^6) + (1\times 2^5) + (0\times 2^4) \\ &\quad + (0\times 2^3) + (0\times 2^2) \\ &\quad + (0\times 2^1) + (1\times 2^0) \\ &= 97\end{aligned}$$

Answer is A.

26. Computer data is in digital format. For it to be communicated over the phone lines using a modem, the digital data must be converted into an analog signal by modulating/demodulating a carrier wave using a shift-keying method, such as frequency shift-keying or phase shift-keying.

Answer is D.

27. The visually appealing world wide web communications technology has revolutionized data communications usage within organizations. Choices (A) and (B) are older developments that are not competitive, and choice (D) is a competing information technology, although on an altogether different platform that is not based on data communications.

Answer is C.

28.

$$\begin{aligned}t &= \text{time} \\ f &= \text{feed} \\ s &= \text{spindle speed} \\ L &= \text{length of cut}\end{aligned}$$

$$\begin{aligned}t &= \frac{L}{fs} = \frac{24 \text{ in}}{\left(0.012 \ \frac{\text{in}}{\text{rev}}\right)\left(400 \ \frac{\text{rev}}{\text{min}}\right)} \\ &= 5.0 \text{ min}\end{aligned}$$

Answer is B.

29.

$$\begin{aligned}t &= \text{time} \\ n &= \text{number of slugs} \\ p &= \text{set-up time} \\ A &= \text{cut area} \\ r &= \text{cutting rate}\end{aligned}$$

$$\begin{aligned}t &= n\left(p + \frac{A}{r}\right) \\ &= (12)\left[(15 \text{ sec})\left(\frac{1 \text{ min}}{60 \text{ sec}}\right) + \frac{\pi\left(\frac{3 \text{ in}}{2}\right)^2}{12 \ \frac{\text{in}^2}{\text{min}}}\right] \\ &= 10.07 \text{ min}\end{aligned}$$

Answer is C.

30. Modern standard beverage cans are made through a series of steps including the deep drawing and stretching of aluminum sheet stock to form the can body. The can tabs are blanked and formed and then attached using a stud rivet to the blanked can top. The can is then filled with the liquid, and the tab/top assembly is crimped to the body. Seam welding, once used with steel cans, is not longer required.

Answer is D.

31. Reducing the number of components, including the use of snap fasteners, is the most appropriate design for assembly/manufacturing guidelines for improving the product as described. Rule 1 does not apply because there are no flexible components (springs, wires, etc.), rule 2 does not apply because the current operations are already as simple as possible, and rule 4 does not apply because there is no problem of confusing insertion geometry in the case of this product.

Answer is C.

32. A sculptured surface model requires a higher-order polynomial equation and will be more precisely represented by a cubic parametric equation such as a nonuniform rational B-spline model for machining purposes. The polyhedral solid model in choice (C) and the $2^1/_2$-dimensional model in choice (B) are only approximation methods that at best will produce a crude result. The prismatic model in choice (A) does not apply to this surface.

Answer is D.

33. Given a flat bottom in a pocket, the correct procedure is to use a flat or regular end mill in a spiral tool path with a left offset from the pocket boundary in a counterclockwise direction. The regular end mill results in a flat pocket bottom. The spiral tool path will take less time than other profiles for a circular boundary. A counterclockwise path requires a left offset. The other three alternatives have one or more errors in them, ranging from ridges left by a ball nose mill to a bigger than desired pocket from the wrong offset relative to the tool direction.

Answer is C.

34. Since parts are palletized 5 per pallet, the arrival rate for pallets to the conveyor is 100/5 or 20 pallets per minute. Since the conveyor is only 1.2 m wide, the pallet must be placed on the conveyor with the 1.3 m side parallel with the conveyor. This implies that a minimum 1.3 m footprint is required for each pallet on the conveyor; therefore, the conveyor must go fast enough to clear a space for each new pallet. The minimum speed the conveyor must go is

$$\left(20\ \frac{\text{loads}}{\text{min}}\right)(1.3 \text{ m per load}) = 26 \text{ m per minute}$$

Answer is C.

35. p = utilization

L = mean arrival rate = 40 moves/hr

s = no. of servers = 4 forklifts

u = mean service time

$$p = \frac{L}{su}$$

$$u = \frac{60\ \frac{\text{min}}{\text{hr}}}{1 \text{ min load} + 1 \text{ min unload} + \frac{240 \text{ m}}{80\ \frac{\text{m}}{\text{min}}}} = 12$$

$$s = \frac{40}{(4)(12)} = 0.8333 \quad (83.33\%) \quad \text{[round to 85\%]}$$

Answer is C.

36. The product 1 distance is

$$50 \text{ m} + 50 \text{ m} = 100 \text{ m}$$

The product 2 distance is

$$100 \text{ m} + 50 \text{ m} + 50 \text{ m} + 100 \text{ m} = 300 \text{ m}$$

$$\begin{aligned}&(\text{total flow})(\text{distance})\\ &\quad = (100 \text{ loads})(100 \text{ m}) + (200 \text{ loads})(300 \text{ m})\\ &\quad = 70{,}000 \text{ load·m}\end{aligned}$$

Answer is D.

37. This is an unconstrained optimization problem and is solved by taking the partial derivative with respect to x and y, setting them equal to zero, and solving the resulting system of equations for the critical point. If the function is then convex at the point, it is a local minimum. For this problem,

$$\frac{\partial f(x,y)}{\partial x} = 4x + 4y - 4 = 0$$

$$\frac{\partial f(x,y)}{\partial y} = 4x + 2y - 2 = 0$$

This results in the following system of equations.

$$\begin{aligned}4x + 4y &= 4\\ 4x + 2y &= 2\end{aligned}$$

This system reduces to

$$\begin{aligned}4x + 4y &= 4\\ -2y &= -2\end{aligned}$$

These equations yield the solution $x = 0$ and $y = 1$. However, the Hessian matrix of second partials is given by

$$H\big(f(x,y)\big) = \begin{pmatrix} \frac{\partial^2 f(x,y)}{\partial x \partial x} & \frac{\partial^2 f(x,y)}{\partial x \partial y} \\ \frac{\partial^2 f(x,y)}{\partial y \partial x} & \frac{\partial^2 f(x,y)}{\partial y \partial y} \end{pmatrix} = \begin{pmatrix} 4 & 4 \\ 4 & 2 \end{pmatrix}$$

$H\big(f(x,y)\big)$ has a determinant of $(4)(2) - (4)(4) = -8$, which means the Hessian matrix is indefinite. This indicates that the point is an inflection point.

Answer is C.

38. This problem can be solved by the Hungarian method for assignment problems by giving the assignments that are not possible a large associated cost, say \$1000. For this problem, the cost matrix is

$$\begin{vmatrix} 20 & 1000 & 20 & 8 \\ 14 & 16 & 8 & 12 \\ 36 & 12 & 20 & 28 \\ 28 & 8 & 24 & 1000 \end{vmatrix}$$

The reduced cost matrix is found by subtracting the smallest value in each row from each of the other elements in that row and then, for the resultant matrix, subtracting the smallest element in each column from each of the other elements in the column. For the above matrix subtract 8 from each element of row one, 8 from each element of row two, 12 from each element of row three, and 8 from each element of row four. The resulting matrix is

$$\begin{vmatrix} 12 & 992 & 12 & 0 \\ 6 & 8 & 0 & 4 \\ 24 & 0 & 8 & 16 \\ 20 & 0 & 16 & 992 \end{vmatrix}$$

For this matrix subtract 6 from each element of column one, and 0 from each element of the other three columns, resulting in the following reduced cost matrix.

$$\begin{vmatrix} 6 & 992 & 12 & 0 \\ 0 & 8 & 0 & 4 \\ 18 & 0 & 8 & 16 \\ 14 & 0 & 16 & 992 \end{vmatrix}$$

The minimum number of lines to cover the zeros for the reduced matrix is three.

$$\begin{vmatrix} 6 & 992 & 12 & 0 \\ 0 & 8 & 0 & 4 \\ 18 & 0 & 8 & 16 \\ 14 & 0 & 16 & 992 \end{vmatrix}$$

The new reduced cost matrix is found by subtracting the smallest uncovered number from each uncovered number and adding that number to each twice covered number. In the previous matrix, 8 is the smallest uncovered number. The new reduced cost matrix is

$$\begin{vmatrix} 6 & 1000 & 12 & 0 \\ 0 & 16 & 0 & 4 \\ 10 & 0 & 0 & 8 \\ 6 & 0 & 8 & 984 \end{vmatrix}$$

The minimum number of lines to cover the zeros for this matrix is four.

$$\begin{vmatrix} 6 & 1000 & 12 & 0 \\ 0 & 16 & 0 & 4 \\ 10 & 0 & 0 & 8 \\ 6 & 0 & 8 & 984 \end{vmatrix}$$

Hence, this matrix gives the optimal solution, which is obtained from the zero cells. Operator 1 must be assigned to machine 4, and operator 4 must be assigned to machine 2. After these assignments, operator 3 must be assigned to machine 3, and lastly operator 2 to machine 1. The optimal assignment is then 1-4, 2-1, 3-3, and 4-2. The associated cost for this assignment from the original cost matrix is

$$8 + 14 + 20 + 8 = 50$$

Answer is C.

39. To find the solution given by the northwest corner rule, the transportation matrix must be formed. There is one row for each source and one column for each destination. There is also a row for demand and a column for supply. To follow the northwest corner rule, start in the upper left corner and assign to that cell as much as possible to meet either the supply or demand for that row or column. If the supply is met, go to the adjacent cell in the column and repeat the process. If the demand is met, go to the adjacent cell in the row and repeat the process. If both the demand and supply are met, put a 0 in either the adjacent column cell or the adjacent row cell. If a 0 is put in the adjacent column cell, move to the adjacent row cell. If the 0 is put in the adjacent row cell, move to the adjacent column cell. Continue until each row and column has its supply and demand met. This is shown as follows.

	destinations					
source	1	2	3	4	5	supply
1	5	15				20
2		0	40	0		40
3				10		10
4				25	35	60
demand	5	15	40	35	35	

5 is put in cell $(1,1)$, meeting the demand of column 1. Then move to the adjacent row cell $(1,2)$ and assign the value of 15. This meets both the supply and demand of row 1 and column 2, respectively. Since both are met, put a 0 in the adjacent column cell $(2,2)$, then move to the row cell $(2,3)$, which is adjacent to $(2,2)$. For this cell, a value of 40 is assigned, which again meets both the supply and demand of the corresponding row and column (i.e., row 2 and column 3). Since both are met, put a 0 in the adjacent row cell $(2,4)$, then move to the

adjacent column cell $(3,4)$. A value of 10 is assigned to this cell, meeting the supply of row 3. Then move to the adjacent column cell $(4,4)$ and assign a value of 25, meeting the demand of column 4. Since the column demand is met, move to the adjacent row cell $(4,5)$ and assign a value 35, meeting the supply and demand for row 4 and column 5, completing the initial solution since all the demands and supplies are now met.

The corresponding objective function value is then

$$\begin{aligned}&(5)(10) + (15)(15) + (0)(4) + (40)(12)\\&+ (0)(23) + (10)(28) + (25)(51) + (35)(29) = 3325\end{aligned}$$

Answer is D.

40. The makespan is determined by the time at which the last job in the sequence finishes at the second station. Starting with the first job in the sequence, determine the start and finish time of the painting operation and the start and finish time of the inspection operation. The next job in the sequence will begin its painting operation immediately following the finish of the previous job in the sequence. A job will begin the inspection operation when its painting operation is complete and the inspection station has finished the preceding job in the sequence. The following table shows the scheduled start and finish times of all jobs at each of the stations.

	painting operation		inspection operation	
	start time	finish time	start time	finish time
job 3	0	15	15	23
job 2	15	23	23	28
job 1	23	35	35	44
job 4	35	42	44	48

Job 4, the last job in the sequence, is scheduled to complete its inspection operation at time 48. Thus, the makespan of the sequence $\{3,2,1,4\}$ is 48.

Answer is D.

41. The general formula for determining the economic order quantity is

$$Q^* = \sqrt{\frac{2K\lambda}{h}}$$

$$\begin{aligned}K &= \text{cost to place an order} = \$15\\ h &= \text{holding cost per unit per period}\\ &= \$0.10/\text{unit-wk}\end{aligned}$$

$$\begin{aligned}\lambda &= \text{average demand per period}\\ &= 750 \text{ units/wk}\end{aligned}$$

$$\begin{aligned}Q^* &= \sqrt{\frac{(2)(\$15)\left(750\ \frac{\text{units}}{\text{wk}}\right)}{\frac{\$0.10}{\text{unit-wk}}}}\\ &= 474.3 \text{ units} \quad (475 \text{ units})\end{aligned}$$

Answer is C.

42. The general equation for determining the minimum workforce requirement is

$$W = \frac{D_w T_p}{S_w H}$$

$$\begin{aligned}W &= \text{minimum number of workers per shift}\\ D_w &= \text{average weekly demand} = 10{,}000 \text{ units/wk}\\ T_p &= \text{production time per unit} = 0.25 \text{ hr/unit}\\ S_w &= \text{number of shifts per week}\\ &= \left(2\ \frac{\text{shifts}}{\text{day}}\right)\left(5\ \frac{\text{days}}{\text{wk}}\right)\\ &= 10 \text{ shifts/wk}\\ H &= \text{number of productive hours per person}\\ &= 8.5\ \frac{\text{hr}}{\text{person}} - 1.5\ \frac{\text{hr}}{\text{person}}\\ &= 7 \text{ hr/person}\end{aligned}$$

$$\begin{aligned}W &= \frac{\left(10{,}000\ \frac{\text{units}}{\text{wk}}\right)\left(0.25\ \frac{\text{hr}}{\text{unit}}\right)}{\left(10\ \frac{\text{shifts}}{\text{wk}}\right)\left(7\ \frac{\text{hr}}{\text{person}}\right)}\\ &= 35.7 \text{ people/shift} \quad (36 \text{ people/shift})\end{aligned}$$

A minimum workforce of 36 people per shift is needed to meet the demand without overtime. Note that if only 35 people per shift are used, weekly production will be less than weekly demand.

Answer is D.

43. Tools for the graphical analysis of worker productivity include operation process charts to analyze if operations are needed, flow process charts to conduct a more detailed analysis of either a product's operations or a worker's processes, and left-hand/right-hand charts to analyze all of the activities that a worker performs.

Answer is D.

44. $$n = \frac{p(1-p)}{\left(\frac{Sp}{F(z)}\right)^2}$$

n = number of observations

p = percent of time an activity occurs

S = degree of accuracy required

$F(z)$ = standard normal distribution variant for 95% confidence

$$n = \frac{(0.2)(1-0.2)}{\left(\frac{(0.05)(0.2)}{1.64}\right)^2} = 4303 \text{ observations}$$

Answer is B.

45.

	year 1	year 2	year 3	year 4	year 5
total variable cost	\$150,000	\$170,000	\$180,000	\$210,000	\$220,000
total units produced	120,000	150,000	155,000	195,000	210,000
variable cost / unit	1.25	1.133	1.161	1.076	1.047
percentage change from previous year		−9.36%	2.47%	−7.32%	−2.70%

$$\frac{\text{variable cost}}{\text{unit}} = \frac{\text{total variable cost}}{\text{total units produced}}$$

$$\text{percentage change for year } n = \frac{\text{VCU}(n) - \text{VCU}(n-1)}{\text{VCU}(n-1)}$$

The average change in VCU (variable cost per unit) over the five years is

$$\frac{-9.36\% + 2.47\% + (-7.32\%) + (-2.70\%)}{4} = -4.23\% \quad (-4\%)$$

Answer is A.

46. This problem can be viewed as a discrete Markov process. To find the expected value of stock 2, the steady-state probabilities of the values of stock 2 must be known. Let state 1 represent \$20 and state 2 represent \$35. The corresponding one-step probability transition matrix for stock 2 is

$$\mathbf{P} = \begin{vmatrix} 0.9 & 0.1 \\ 0.2 & 0.8 \end{vmatrix}$$

To find the steady-state probabilities π, solve the system of equations given by $\boldsymbol{\pi}\mathbf{P} = \boldsymbol{\pi}$, which is equivalent to $\boldsymbol{\pi}(\mathbf{P} - \mathbf{I}) = \mathbf{0}$. This system can be written as

$$(\pi_1, \pi_2)\begin{pmatrix} -0.1 & 0.1 \\ 0.2 & -0.2 \end{pmatrix} = (0, 0)$$

This yields the following system of equations.

$$-0.1\pi_1 + 0.2\pi_2 0$$
$$0.1\pi_1 - 0.2\pi_2 = 0$$

This system reduces to $-0.1\pi_1 + 0.2\pi_2 = 0$, which has the solution $\pi_1 = 2\pi_2$. The normalizing equation $(\Sigma\pi_i = 1)$ yields $2\pi_2 + \pi_2 = 1$, which implies $\pi_2 = \frac{1}{3} = 0.3333$ and $\pi_1 = \frac{2}{3} = 0.6666$. The expected value of the stock is then given by

$$\left(\tfrac{2}{3}\right)(\$20) + \left(\tfrac{1}{3}\right)(\$35) = \$25$$

Answer is B.

47. Since the arrivals occur according to a Poisson process, the interarrival times are exponential (M). Even though the service (machining) time is exponential for each machine, service times are not exponential because the service occurs in three consecutive stages. Therefore, service is actually the total of the three stages. But since the stages are consecutive and each stage has exponential service times, service times are distributed according to an Erlang-3 distribution (E_3). The capacity of the buffer is the limiting factor on the number of products in the system. Hence the capacity of the total number of products in the system is 11:1 being machined and 10 waiting.

Answer is D.

48. Interarrival times and service times are exponentially distributed. There is also room for a large number of people in the food court, and people can wait outside the court, implying that the capacity of the system is infinite (very large). Therefore, this is an M/M/2 model. For this model, the number of people waiting in line is given by

$$L_q = \frac{2\rho^3}{1-\rho^2}$$

For this problem, $\lambda = 30/\text{hr}$ is the arrival rate and $\mu = 20/\text{hr}$ is the service rate for each teller.

The server utilization is

$$\rho = \frac{\lambda}{s\mu} = \frac{\frac{30}{\text{hr}}}{(2)\left(\frac{20}{\text{hr}}\right)} = 0.75$$

Therefore, the number of people waiting in line is

$$L_q = \frac{(2)(0.75)^3}{1-(0.75)^2} = 1.93 \text{ people} \quad (2 \text{ people})$$

Answer is C.

49. The central limit theorem states that when a series of independent means (or sums) are combined into a single mean (or sum), the resulting mean (or sum) has a distribution that approaches normal as the number of samples being combined increases. Therefore, the resultant statistic can be assumed to come from an independent, identical, normally distributed data set. These assumptions are required to apply most standard statistical tests such as confidence intervals, tests of equal means, and so on.

Answer is B.

50. The central limit theorem states that when a series of independent means (or sums) are combined into a single mean (or sum), the resulting mean (or sum) has a distribution that approaches normal as the number of samples being combined increases. Therefore, the resultant statistic can be assumed to come from an independent, identical, normally distributed data set. These assumptions are required to apply most standard statistical tests such as confidence intervals, tests of equal means, and so on.

Answer is B.

51. A 95% confidence interval on mean transporter utilization says that there is a 95% chance the population (or true) mean transporter utilization lies within the given interval.

Answer is B.

52. Use the Poisson distribution to approximate the binomial probabilities.

$$P(\text{accept lot}|P, n, c) = P(x=0, n=3, nP=3P)$$

$$P = \text{population process proportion defective}$$

$$\begin{aligned} P(x=0) &= \frac{e^{-nP}(nP)^x}{x!} = \frac{e^{-(3)(0.30)}\left((3)(0.30)\right)^0}{0!} \\ &= 0.407 \quad (40\%) \end{aligned}$$

Answer is C.

53. $\overline{R}$ average of sample ranges $= \Sigma R/K$

K = number of sample ranges = 25

d_2 = constant factor for sample size $n = 5$

d_2 relates the average ranges to an estimate of the population standard deviation $\hat{\sigma}_X$.

$$\hat{\sigma}_X \approx \frac{\overline{R}}{d_2} = \frac{8.960}{2.326} = 3.852 \quad (3.850)$$

Answer is A.

54. D_4 is a constant factor to approximate the upper range control limit three standard deviations above the R-chart central line.

D_3 is a constant factor to approximate the lower range control limit three standard deviations below the R-chart central line.

$$\begin{aligned} \text{UCL}_R &= D_4\overline{R} = (2.114)(8.960) = 18.941 \\ \text{LCL}_R &= D_3\overline{R} = (0)(8.960) = 0 \end{aligned}$$

Answer is B.

55. "Organizations that practice employee empowerment don't just allow employee empowerment, they actively seek it. Empowered employees provide input concerning decisions that affect them and can supply their own ingenuity in seeking improvement themselves within specified limits." (Goetoch, D.L., and Davis, S.B., *Introduction to Total Quality*, Prentice-Hall, second edition, 1997 (pp 205–6).)

The implication is that motivated employees will more than recover costs of training in their subsequent proposed improvements.

Answer is C.

56. "The most prestigious award for quality in the United States is the Malcolm-Baldrige National Quality Award. Competitors . . . are evaluated according to seven categories." (Goetoch, D.L., and Davis, S.B., *Introduction to Total Quality*, Prentice-Hall, second edition, 1997 (pp 437).)

Answer is C.

57. "A quality circle is a group of employees that voluntarily meets regularly for the purpose of identifying, recommending, and making workplace improvements." (Goetoch, D.L., and Davis, S.B., *Introduction to Total Quality*, Prentice-Hall, second edition, 1997 (pp 189).)

Answer is D.

58. The general formula for finding the sample size required for an activity with a stated accuracy and a confidence of 95% is

$$n = \frac{(1.96)^2(p)(1-p)}{\epsilon^2}$$

n = required sample size for the activity
p = proportion activity observed in the pilot study
1.96 = standardized normal variant for 95% confidence
ϵ = desired accuracy

Machine idle—no product:

$$p = \frac{5}{100} = 0.05$$

$$n = \frac{(1.96)^2(0.05)(0.95)}{(0.05)^2} = 72.99 \quad (73)$$

Machine operating:

$$p = \frac{85}{100} = 0.85$$

$$n = \frac{(1.96)^2(0.85)(0.15)}{(0.05)^2} = 195.9 \quad (196)$$

Machine idle—downtime and repair:

$$p = \frac{10}{100} = 0.10$$

$$n = \frac{(1.96)^2(0.10)(0.90)}{(0.05)^2} = 138.3 \quad (139)$$

The largest number of required samples is 196.

$$\begin{aligned} &196 \text{ observations required} - 100 \text{ current observations} \\ &\quad = 96 \text{ additional observations} \end{aligned}$$

Answer is C.

59. The general formula for allowed time before allowances is

$$\text{AT} = (\text{OT})\left(\frac{R}{100}\right)K$$

OT = observed time
R = objective rating as a percent
K = occurrences per cycle

For element 1,

$$\text{AT} = (0.246)\left(\frac{110}{100}\right)(1) = 0.2706$$

For element 2,

$$\text{AT} = (1.214)\left(\frac{100}{100}\right)(1) = 1.2140$$

For element 3,

$$\text{AT} = (0.252)\left(\frac{110}{100}\right)(1) = 0.2772$$

For element 4,

$$\text{AT} = (1.682)\left(\frac{90}{100}\right)\left(\frac{1}{3}\right) = 0.5046$$

The total for the job is

$$0.2706 + 1.2140 + 0.2772 + 0.5046 = 2.2664$$

Answer is B.

60. The general formula for determining the number of cycles to time per element for any element that is not machine-controlled is

$$n = \left(\frac{1.96s}{\epsilon}\right)^2$$

1.96 = normal standardized variant for 95% accuracy
s = element standard deviation
ϵ = accuracy

For element 1,

$$n = \frac{(1.96)^2(0.031)^2}{(0.01)^2} = 36.9 \quad (37)$$

Element 2 is machine controlled.

For element 3,

$$n = \frac{(1.96)^2(0.022)^2}{(0.01)^2} = 18.6 \quad (19)$$

For element 4,

$$n = \frac{(1.96)^2(0.015)^2}{(0.01)^2} = 8.64 \quad (9)$$

There are sufficient samples for elements 2, 3 and 4, but element 1 requires seven additional observations. The job must be timed for an additional seven cycles.

Answer is B.

More FE/EIT Exam Practice!

All FE/EIT examinees take the same General test in the morning session of the FE exam. To prepare for this test, you can use any of the following publications, depending on how much and what kind of review you need.

FE Review Manual: Rapid Preparation for the General Fundamentals of Engineering Exam

Michael R. Lindeburg, PE

The *FE Review Manual* prepares you for the FE/EIT exam in the fastest, most efficient way possible. Completely up-to-date for the exam content and format, this book gives you diagnostic tests to see what you need to study most, concise reviews of all exam topics, 1150+ practice problems with solutions, and a complete eight-hour sample exam with solutions. The *FE Review Manual* is your best choice if you are still in college or a recent graduate, or if your study time is limited.

999 Nonquantitative Problems for FE Examination Review

Kenton Whitehead, PhD, PE

Passing the FE exam means answering questions quickly and accurately. Nonquantitative problems on the exam don't require numerical calculations but rather an understanding of theory and principle. It's essential that you answer these questions quickly, leaving yourself more time to work on quantitative problems. This book will bring you up to speed on the concepts you need to know and give you intensive practice with problems you won't find anywhere else. Answers are included.

FE/EIT Sample Examinations

Michael R. Lindeburg, PE

Testing yourself with a realistic simulation of the FE exam is an essential part of your preparation. There's no better way to get ready for the time pressure of the exam. *FE/EIT Sample Examinations* includes two complete eight-hour practice tests in multiple-choice format. Fully worked-out solutions are included. Use these sample exams to build skills and confidence for taking the general FE exam.

Calculus Refresher for the Fundamentals of Engineering Exam

Peter Schiavone, PhD

Many engineers report having more trouble with problems involving calculus than with anything else on the FE/EIT exam. This book covers all the areas that you'll need to know for the exam: differential and integral calculus, centroids and moments of inertia, differential equations, and precalculus topics such as quadratic equations and trigonometry. You get clear explanations of theory, relevant examples, and FE-style practice problems (with solutions). If you are at all unsure of your calculus skills, this book is a must.

Professional Publications, Inc.
1250 Fifth Avenue • Belmont, CA 94002
(800) 426-1178 • Fax (650) 592-4519
For secure, convenient ordering, try our complete online catalog at
www.ppi2pass.com

Source Code BOC ☞ **Quick — *I need additional PPI study materials!***

Please send me the PPI exam review products checked below. I have provided my credit card number and authorize you to charge your current prices, plus shipping, to my card.

For the FE Exam

- ☐ FE Review Manual
- ☐ Civil Discipline-Specific Review for the FE/EIT Exam
- ☐ Mechanical Discipline-Specific Review for the FE/EIT Exam
- ☐ Electrical Discipline-Specific Review for the FE/EIT Exam
- ☐ Chemical Discipline-Specific Review for the FE/EIT Exam
- ☐ Industrial Discipline-Specific Review for the FE/EIT Exam
- ☐ Engineer-In-Training Reference Manual ☐ Solutions Manual, SI Units

For the PE, SE, and PLS Exam

- ☐ Civil Engrg Reference Manual ☐ Practice Problems
- ☐ Mechanical Engrg Reference Manual ☐ Practice Problems
- ☐ Electrical Engrg Reference Manual ☐ Practice Problems
- ☐ Environmental Engrg Reference Manual ☐ Practice Problems
- ☐ Chemical Engrg Reference Manual ☐ Practice Problems
- ☐ Structural Engrg Reference Manual
- ☐ PLS Sample Exam

These are just a few of the products we offer for the FE, PE, SE, and LS exams. For a full list, visit our website at www.ppi2pass.com.

For fastest service,
Web **www.ppi2pass.com**
Call **800-426-1178** Fax **650-592-4519**

Mail this form to:
PPI, 1250 Fifth Ave., Belmont, CA 94002

NAME/COMPANY ______________________

STREET ______________________ SUITE/APT ________

CITY ______________ STATE ________ ZIP ________

DAYTIME PH # ______________ EMAIL ______________

VISA/MC/DSCVR # ______________________ EXP. DATE ________

CARDHOLDER'S NAME ______________________

SIGNATURE ______________________

Email Updates Keep You on Top of Your Exam

You need current information to be fully prepared for your exam. Register for PPI's Email Updates to receive convenient updates relevant to the specific exam you are taking. Our updates include notices of exam changes, useful exam tips, errata postings, and new product announcements. There is no charge for this service, and you can cancel at any time.

Register at **www.ppi2pass.com/cgi-bin/signup.cgi**

Free Catalog of Tried-and-True Exam Products

Get a free PPI catalog with a comprehensive selection of the best FE, PE, SE, FLS, and PLS exam-review products available, user tested by more than 800,000 engineers and surveyors. Included are books, software, videos, and the NCEES sample-question books.

Request a catalog at **www.ppi2pass.com/catalogrequest**

How to Report Errors

Find an error? You can report it in two easy steps.

First, check the errata listings on our website, at **www.ppi2pass.com/errata**. The item you noticed may already have been identified. It's always a good idea to check this page before you start studying and periodically thereafter.

Then, go to PPI's Errata Report Form at **www.ppi2pass.com/erratasubmit**, and tell us about the discrepancy you think you've found. Your information will be forwarded to the appropriate author or subject matter expert for verification. Valid corrections are added to the errata section of our website.

You may also fax errata to us at 650-592-4519 or mail them to Professional Publications, Inc., c/o Editorial Errata Department, 1250 Fifth Ave., Belmont, CA 94002.
Be sure to include your name, the book title, the edition and printing numbers, the page number(s), and any other information that will help us locate the error(s).